Norton Commando Owners Workshop Manual

by Jeff Clew

(with an additional chapter on the electric start models by Stewart Wilkins)

Models Covered:

Commando. 745cc. February 1968 to March 1969
Commando Fastback. 745cc. March 1969 to March 1973
Commando Roadster. 745cc. March 1970 to October 1973
Commando Hi-Rider. 745cc. May 1971 to October 1973
Commando Interstate. 745cc. January 1972 to October 1973
Commando 850 Roadster. 828cc. April 1973 to October 1977
Commando 850 Hi-Rider. 828cc. April 1973 to February 1975
Commando 850 Interstate. 828cc. April 1973 to October 1977

ISBN 978 0 85696 125 0

ABCDE
FGHIJ
KLMNO
PQRS
3

Printed in India   *(125-6P9)*

Haynes Group Limited
Sparkford, Yeovil,
Somerset BA22 7JJ, England

Haynes North America, Inc
2801 Townsgate Road,
Suite 340, Thousand Oaks,
CA 91361, USA

Acknowledgements

Our grateful thanks are due to Norton-Villiers Limited for technical assistance given and permission to use many of their illustrations. Brian Horsfall gave necessary assistance with the overhaul and devised ingenious methods for overcoming the lack of service tools. Les Brazier took the photographs that accompany the text. Tim Parker edited the text.

Our thanks are also due to Steve Challis who kindly loaned us the 750 cc Commando model on which this manual is based and to Arthur Vincent of Vincent and Jerrom, the Norton-Villiers dealer in Taunton, who provided much in-depth servicing and modification information. We are also grateful to Fran

Ridewood and Co, of Wells, who provided the Mark 3 Electric Start model used for the photographic sequences in Chapter 9, and to Chris Rogers who provided the 850 Roadster featured on the front cover of this manual.

Finally, we would also like to acknowledge the help of the Avon Rubber Company who kindly supplied illustrations and advice about tyre fitting, of Amal Limited for the use of their carburettor illustrations, and of Automotive Products Limited who provided illustrations and advice about their hydraulic disc brakes.

About this manual

The author of this manual has the conviction that the only way in which a meaningful and easy to follow text can be written is first to do the work himself, under conditions similar to those found in the average household. As a result, the hands seen in the photographs are those of the author. Even the machines are not new; examples that have covered a considerable mileage were selected, so that the conditions encountered would be typical of those found by the average owner/rider. Unless specially mentioned and therefore considered essential, Norton-Villiers service tools have not been used. There is invariably alternative means of loosening or slackening some vital component, when service tools are not available and risk of damage is to be avoided at all costs.

Each of the chapters is divided into numbered sections. Within the sections are numbered paragraphs. Cross-reference throughout this manual is quite straightforward and logical. When reference is made, 'See section 6.10', it means section 6, paragraph 10 in the same chapter. If another chapter were meant

it would say, 'See Chapter 2, Section 6.10'.

All photographs are captioned with a section/paragraph number to which they refer, and are always relevant to the chapter text adjacent.

Figure numbers (usually line illustrations) appear in numerical order, within a given chapter. 'Fig 1.1' therefore refers to the first figure in chapter 1.

Left hand and right hand descriptions of the machines and their components refer to the left and right of a given machine when normally seated.

Motorcycle manufacturers continually make changes to specifications and recommendations, and these, when notified, are incorporated into our manuals at the earliest opportunity.

Whilst every care is taken to ensure that the information in this manual is correct no liability can be accepted by the authors or publishers for loss, damage or injury caused by any errors in or omissions from the information given.

Contents

TERRY DAVEY ©HAYNES H.II

Model dimensions:

	Fastback	Roadster	Interstate	Hi-Rider	850 cc
Height:	40.75 in (103.5 cm)	40.75 in (103.5 cm)	40.75 in (103.5 cm)	50.25 in (127.6 cm)	40.75 in (103.5 cm)
Length:	87.5 in (222 cm)	87.5 in (222 cm)	87.5 in (222 cm)	87.5 in (222 cm)	88.0 in (223.7 cm)
Width:	26.0 in (66 cm)	26.0 in (66 cm)	26.0 in (66 cm)	26.0 in (66 cm)	26.0 in (66 cm)
Ground clearance:	6.0 in (15 cm)	6.0 in (15 cm)	6.0 in (15 cm)	6.0 in (15 cm)	6.0 in (15 cm)
Wheelbase:	56.75 in (144 cm)	56.75 in (144 cm)	56.75 in (144 cm)	56.75 in (144 cm)	58.0 in (147cm)
Weight (dry):	395.4 lb (197.3 kg)	395.4 lb (179.3 kg)	410 lb (186 kg)	N/A	418 - 430 lb* (190 - 196 kg) *

* Depending on specification

Introduction to the Norton Commando

In the immediate post-war years, British motor cycle manufacturers began to appreciate the advantages of a vertical twin engine, acknowledging the success of one manufacturer who had succeeded in marketing a highly attractive design as far back as 1937. The first Norton twin was unveiled at the 1949 Earl's Court Show. Of 497 cc capacity it was marketed as the Model 7. The specification included a cast iron engine with splayed exhaust ports and pushrods located in cast-in tunnels at the front of the cylinder block. The magneto and camshaft were driven by separate chains that formed part of the timing gear assembly and were fully enclosed within a conventional timing cover. Norton Roadholder forks looked after the front suspension; at the rear end a 'garden gate' frame layout employed plunger units characteristic of that era. As time progressed, the Norton twins assumed the name Dominator and the overall specification was changed to include alloy engine components, the famous Norton 'Featherbed' frame with swinging arm rear suspension and shortened Roadholder forks. Machines fitted with the 'Featherbed' frame were redesignated the Model 88 and during 1956 a 596 cc version, known as the Model 99, was added to the range. Sports versions of both models were introduced during 1962 and a 646 cc machine to the same basic specification that was available in both standard and sports trim.

Financial problems necessitated Norton Motors becoming part of the Associated Motor Cycles Group and in 1962 the old Bracebridge Street works in Birmingham, the traditional home of the Norton, finally closed when all production was moved to Woolwich. Although the Norton range of motor cycles was continued, some hybrids were marketed comprising in several instances Norton engines fitted into AMC cycle parts. The twins were, however, the least affected, apart from some export only versions. It was about this time that the Norton Atlas came into being, fitted with an enlarged twin cylinder engine of 745 cc. It too ultimately acquired an AMC frame. By now the British motor cycle industry in general was faced with financial problems, mainly on account of the Japanese invasion of their traditional markets. Associated Motor Cycles were one of several manufacturers to collapse under the strain and towards the end of 1966 a satisfactory arrangement was completed for their assets to be acquired by Manganese Bronze Holdings Limited, of which Villiers Engineering was a subsidiary. A new

company was born - Norton-Villiers Limited and work commenced on a design project under the direction of Dr Stefan Bauer at the Villiers premises in Wolverhampton. Meanwhile, Norton production was again on the move, this time to new premises on one of the trading estates at Andover. The move was completed during 1969. The outcome of Dr Bauer's project was the Norton Commando, a 750 cc vertical twin using an inclined version of the original Dominator engine suspended in an entirely new frame by means of Isolastic suspension mountings, a patented Norton-Villiers invention. Although initially dismissed by the sceptics as merely a revamped Atlas, the Commando soon found favour with all who were privileged to have a test ride. The Isolastic suspension eliminated the high frequency vibration normally associated with the vertical twin engine and gave a hitherto unknown smooth ride. From this moment the Commando became a success, as underlined by the presentation of the 'Motor Cycle of the Year' award on no less than five consecutive years as the result of a reader contest organised by Motor Cycle News. Six versions of the Commando have been marketed, the Interstate, designed for long distance riding, the Roadster for normal road use, the Fastback, virtually the original version, the Hi-rider, built in semi-chopper style, the 750 cc Production racer and the Interpol, a fully-equipped police model used by many police forces throughout the world. Only the Fastback is no longer in production. Further permutations are available from the use of the standard or the Combat engine, the latter being a specially-tuned version that develops 65 bhp at 7000 rpm - 5 bhp more than the standard engine.

During March 1973 an 850 cc model was added to the range, fitted with a modified form of the 750 cc engine that has the same power output but imposes less stress on the more vulnerable components. This too is currently available in five versions, one of which is specially equipped to give a significant reduction in noise level.

Perhaps the greatest success of the Commando was Peter Williams' Isle of Man TT win in the 1973 Formula 750 race, when a British manufacturer showed that a British motor cycle could still offer a serious challenge and emerge victorious. The Norton Commando is indeed a superbike and it is particularly fitting that this descriptive attribute was applied to the Commando when interest in large capacity high-performance models heightened several years ago.

Modifications to the Norton Commando range

When the Norton Commando was announced towards the end of 1967, it was greeted with a certain amount of scepticism because on first sight it appeared to comprise of the old Norton Dominator twin cylinder engine mounted at an inclined angle in a set of new cycle parts. It was not realised that the new Isolastic method of engine suspension damped out all engine vibration and produced a machine which had an uncanny smoothness for a vertical twin. In due course the critics were silenced and the Commando had the distinction of being regarded as the first of today's so-called 'Superbikes'.

There can be little doubt that the original design concept has proved correct, since comparatively few modifications of any real consequence have been made since production commenced during 1968. Mostly they relate to styling and the introduction

of 'slim line' front forks. In March 1973 the 750 cc range was supplemented by the introduction of an 850 cc model, the engine of which embodies several design modifications to enable the same power output as that of the 750 cc Combat engine to be obtained, with less stress on the engine components. Today, five variations of the basic 750 cc model are available, together with five versions of the 850 cc model. These include the Interpol model which is used by police forces throughout the world.

Some of the engine and gearbox data contained in this manual can be applied to the earlier Dominator twins of 497 cc and 597 cc, also to the 646 cc twins and the 745 cc Atlas model. It is mainly the cycle parts of these latter models which differ.

Ordering spare parts

When wishing to purchase spare parts for the Norton Commando it is best to deal direct with an accredited Norton-Villiers agent. Parts cannot be obtained direct from Norton-Villiers Limited; all orders must be routed through an approved agent, even if the agent does not hold the parts required in stock. When ordering parts, always quote the frame and engine numbers in full, without omitting any prefixes or suffixes. It is also advisable to include note of the colour scheme of the machine if the parts ordered are required to match in.

The engine number is stamped on the left hand side of the crankcase immediately below the base of the cylinder barrel.

The frame number is stamped on an identification plate rivetted to the steering head.

Always fit parts of genuine Norton-Villiers manufacture. Pattern parts may be available sometimes at lower cost, but they do not necessarily make a satisfactory replacement for the originals. There are cases where reduced life or sudden failure has occurred, to the overall detriment of performance and perhaps rider safety.

Some of the more expendable parts such as spark plugs, bulbs, tyres, oils and greases etc., can be obtained from accessory shops and motor factors, who have convenient opening hours, charge lower prices and can often be found not far from home. It is also possible to obtain parts on a Mail Order basis from a number of specialists who advertise regularly in the motor cycle magazines.

Engine number location

Frame number location

Routine maintenance

Periodic routine maintenance is a continuous process that commences immediately the machine is used. It must be carried out at specified mileage recordings or on a calendar basis if the machine is not used frequently, whichever falls soonest. Maintenance should be regarded as an insurance policy, to help keep the machine in the peak of condition and to ensure long, trouble-free service. It has the additional benefit of giving early warning of any faults that may develop and will act as a regular safety check, to the obvious advantage of both rider and machine alike.

The various maintenance tasks are described under their respective mileage and calendar headings. Accompanying diagrams are provided, where necessary. It should be remembered that the interval between the various maintenance tasks serves only as a guide. As the machine gets older or is used under particularly adverse conditions, it would be advisable to reduce the period between each check.

Some of the tasks are described in detail, where they are not mentioned fully as a routine maintenance item in the text. If a specific item is mentioned but not described in detail, it will be covered fully in the appropriate Chapter. No special tools are required for the normal routine maintenance tasks. The tools contained in the kit supplied with every new machine will prove adequate for each task or if they are not available, the tools found in the average household.

Weekly or every 250 miles

Check oil tank level and replenish if necessary with engine oil of the recommended grade.

Grease brake pedal pivot and cable, oil all other exposed control cables and joints.

Check the battery electrolyte level, chain adjustments and tyre pressures.

Monthly or every 1000 miles

Complete the maintenance tasks listed under the preceding weekly heading, then the following additional items:

Check and, if necessary, adjust both brakes (disc brake is self-adjusting). Make sure disc brake fluid level is correct in reservoir. Examine the disc brake pads for wear.

Check primary chaincase oil level and top up if necessary. Check nuts and bolts for tightness.

Three monthly or every 2500 miles

Complete all the checks listed under the weekly and monthly headings, then the following items:

Drain the oil tank whilst the oil is warm and remove and clean the filters before refilling with new oil of the correct viscosity.

Remove and replace the oil filter element (fitted to all models, 1972 onwards).

Check the gearbox oil level and top up if necessary.

Remove and clean both spark plugs and reset the gaps.

Check the ignition timing after adjusting the contact breaker gaps. It is preferable to use a stroboscope for this check.

Remove and lubricate the final drive chain; change the primary chaincase oil. Check the primary chain adjustment and the clutch adjustment.

Finally, check the Isolastic engine mountings for excess play.

Six monthly or every 5000 miles

Again, complete all the maintenance tasks listed previously, then complete the following additional tasks:

Change the gearbox oil, and also the oil in the front forks.

Check and adjust the camshaft chain.

Clean the contact breaker points and lubricate the contact breaker cam and the auto-advance unit.

Grease the brake operating arm pivots (grease sparingly to prevent grease reaching the brake linings).

Check and, if necessary, adjust the valve clearances.

Fit a new air filter element; dismantle and clean both carburettors. Check that they are synchronised correctly.

Check the swinging arm bushes for play and fill the pivot housing with oil.

Yearly or every 10,000 miles

After completing the weekly, monthly, three-monthly and six-monthly tasks, continue with the following additional items:

Remove and repack the wheel bearings with grease, not omitting the bearing in the centre of the final drive sprocket. Check the primary and secondary chains and sprockets for wear, also both carburettors. If performance has fallen off, decarbonise the engine and regrind the valves.

It should be noted that no special mention has been made relating to the lighting equipment, horn and speedometer, which must be in good working order if the statutory requirements of the UK are to be met. Regulations also apply to the minimum depth of tyre tread and the overall condition of the tyres. It is assumed that every owner/rider will keep a watchful eye on these additional points, especially since they have a direct bearing on rider safety. Remember there is no stage at any point in the life of the machine when a routine maintenance task can be ignored or safety checks neglected.

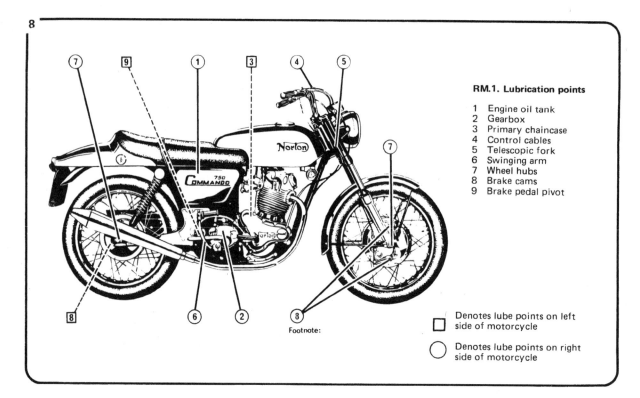

RM.1. Lubrication points

1 Engine oil tank
2 Gearbox
3 Primary chaincase
4 Control cables
5 Telescopic fork
6 Swinging arm
7 Wheel hubs
8 Brake cams
9 Brake pedal pivot

Footnote:

☐ Denotes lube points on left side of motorcycle

◯ Denotes lube points on right side of motorcycle

Recommended lubricants

	Specification	Castrol Product
Engine and primary chaincase	20W/50	Castrol GTX
Gearbox	90 EP	Castrol Hypoy
Swinging arm bushes	140 EP	Castrol Hi-Press
Hubs and cycle parts		Castrol LM Grease
Front forks	10W/30	Castrolite
Rear chain		Castrol Graphited Grease
Hydraulic disc brake		Castrol Girling Universal Brake and Clutch Fluid

Routine maintenance and capacities data

Oil tank	5 Imp pints (6 US pints/2.8 litres)
Gearbox	0.75 Imp pints (0.9 US pints/0.42 litres)
Primary chaincase	7 fl oz (200 cc)
Front forks	5 fl oz (150 cc)
Contact breaker gap	0.014 - 0.016 in (0.35 - 0.04 mm)
Spark plug gap	0.023 - 0.028 in (0.59 - 0.72 mm)
Tappet clearances (engine cold)	0.006 in (0.15 mm) inlet (Commando)
	0.008 in (0.20 mm) inlet (Combat)
	0.008 in (0.20 mm) exhaust (Commando)
	0.010 in (0.25 mm) exhaust (Combat)

Safety first!

Professional motor mechanics are trained in safe working procedures. However enthusiastic you may be about getting on with the job in hand, do take the time to ensure that your safety is not put at risk. A moment's lack of attention can result in an accident, as can failure to observe certain elementary precautions.

There will always be new ways of having accidents, and the following points do not pretend to be a comprehensive list of all dangers; they are intended rather to make you aware of the risks and to encourage a safety-conscious approach to all work you carry out on your vehicle.

Essential DOs and DON'Ts

DON'T start the engine without first ascertaining that the transmission is in neutral.

DON'T suddenly remove the filler cap from a hot cooling system – cover it with a cloth and release the pressure gradually first, or you may get scalded by escaping coolant.

DON'T attempt to drain oil until you are sure it has cooled sufficiently to avoid scalding you.

DON'T grasp any part of the engine, exhaust or silencer without first ascertaining that it is sufficiently cool to avoid burning you.

DON'T allow brake fluid or antifreeze to contact the machine's paintwork or plastic components.

DON'T syphon toxic liquids such as fuel, brake fluid or antifreeze by mouth, or allow them to remain on your skin.

DON'T inhale dust – it may be injurious to health (see *Asbestos* heading).

DON'T allow any spilt oil or grease to remain on the floor – wipe it up straight away, before someone slips on it.

DON'T use ill-fitting spanners or other tools which may slip and cause injury.

DON'T attempt to lift a heavy component which may be beyond your capability – get assistance.

DON'T rush to finish a job, or take unverified short cuts.

DON'T allow children or animals in or around an unattended vehicle.

DON'T inflate a tyre to a pressure above the recommended maximum. Apart from overstressing the carcase and wheel rim, in extreme cases the tyre may blow off forcibly.

DO ensure that the machine is supported securely at all times. This is especially important when the machine is blocked up to aid wheel or fork removal.

DO take care when attempting to slacken a stubborn nut or bolt. It is generally better to pull on a spanner, rather than push, so that if slippage occurs you fall away from the machine rather than on to it.

DO wear eye protection when using power tools such as drill, sander, bench grinder etc.

DO use a barrier cream on your hands prior to undertaking dirty jobs – it will protect your skin from infection as well as making the dirt easier to remove afterwards; but make sure your hands aren't left slippery. Note that long-term contact with used engine oil can be a health hazard.

DO keep loose clothing (cuffs, tie etc) and long hair well out of the way of moving mechanical parts.

DO remove rings, wristwatch etc, before working on the vehicle – especially the electrical system.

DO keep your work area tidy – it is only too easy to fall over articles left lying around.

DO exercise caution when compressing springs for removal or installation. Ensure that the tension is applied and released in a controlled manner, using suitable tools which preclude the possibility of the spring escaping violently.

DO ensure that any lifting tackle used has a safe working load rating adequate for the job.

DO get someone to check periodically that all is well, when working alone on the vehicle.

DO carry out work in a logical sequence and check that everything is correctly assembled and tightened afterwards.

DO remember that your vehicle's safety affects that of yourself and others. If in doubt on any point, get specialist advice.

IF, in spite of following these precautions, you are unfortunate enough to injure yourself, seek medical attention as soon as possible.

Asbestos

Certain friction, insulating, sealing, and other products – such as brake linings, clutch linings, gaskets, etc – contain asbestos. *Extreme care must be taken to avoid inhalation of dust from such products since it is hazardous to health.* If in doubt, assume that they *do* contain asbestos.

Fire

Remember at all times that petrol (gasoline) is highly flammable. Never smoke, or have any kind of naked flame around, when working on the vehicle. But the risk does not end there – a spark caused by an electrical short-circuit, by two metal surfaces contacting each other, by careless use of tools, or even by static electricity built up in your body under certain conditions, can ignite petrol vapour, which in a confined space is highly explosive.

Always disconnect the battery earth (ground) terminal before working on any part of the fuel or electrical system, and never risk spilling fuel on to a hot engine or exhaust.

It is recommended that a fire extinguisher of a type suitable for fuel and electrical fires is kept handy in the garage or workplace at all times. Never try to extinguish a fuel or electrical fire with water.

Note: *Any reference to a 'torch' appearing in this manual should always be taken to mean a hand-held battery-operated electric lamp or flashlight. It does **not** mean a welding/gas torch or blowlamp.*

Fumes

Certain fumes are highly toxic and can quickly cause unconsciousness and even death if inhaled to any extent. Petrol (gasoline) vapour comes into this category, as do the vapours from certain solvents such as trichloroethylene. Any draining or pouring of such volatile fluids should be done in a well ventilated area.

When using cleaning fluids and solvents, read the instructions carefully. Never use materials from unmarked containers – they may give off poisonous vapours.

Never run the engine of a motor vehicle in an enclosed space such as a garage. Exhaust fumes contain carbon monoxide which is extremely poisonous; if you need to run the engine, always do so in the open air or at least have the rear of the vehicle outside the workplace.

The battery

Never cause a spark, or allow a naked light, near the vehicle's battery. It will normally be giving off a certain amount of hydrogen gas, which is highly explosive.

Always disconnect the battery earth (ground) terminal before working on the fuel or electrical systems.

If possible, loosen the filler plugs or cover when charging the battery from an external source. Do not charge at an excessive rate or the battery may burst.

Take care when topping up and when carrying the battery. The acid electrolyte, even when diluted, is very corrosive and should not be allowed to contact the eyes or skin.

If you ever need to prepare electrolyte yourself, always add the acid slowly to the water, and never the other way round. Protect against splashes by wearing rubber gloves and goggles.

Mains electricity and electrical equipment

When using an electric power tool, inspection light etc, always ensure that the appliance is correctly connected to its plug and that, where necessary, it is properly earthed (grounded). Do not use such appliances in damp conditions and, again, beware of creating a spark or applying excessive heat in the vicinity of fuel or fuel vapour. Also ensure that the appliances meet the relevant national safety standards.

Ignition HT voltage

A severe electric shock can result from touching certain parts of the ignition system, such as the HT leads, when the engine is running or being cranked, particularly if components are damp or the insulation is defective. Where an electronic ignition system is fitted, the HT voltage is much higher and could prove fatal.

English/American terminology

Because this book has been written in England, British English component names, phrases and spellings have been used throughout. American English usage is quite often different and whereas normally no confusion should occur, a list of equivalent terminology is given below.

English	American	English	American
Air filter	Air cleaner	Number plate	License plate
Alignment (headlamp)	Aim	Output or layshaft	Countershaft
Allen screw/key	Socket screw/wrench	Panniers	Side cases
Anticlockwise	Counterclockwise	Paraffin	Kerosene
Bottom/top gear	Low/high gear	Petrol	Gasoline
Bottom/top yoke	Bottom/top triple clamp	Petrol/fuel tank	Gas tank
Bush	Bushing	Pinking	Pinging
Carburettor	Carburetor	Rear suspension unit	Rear shock absorber
Catch	Latch	Rocker cover	Valve cover
Circlip	Snap ring	Selector	Shifter
Clutch drum	Clutch housing	Self-locking pliers	Vise-grips
Dip switch	Dimmer switch	Side or parking lamp	Parking or auxiliary light
Disulphide	Disulfide	Side or prop stand	Kick stand
Dynamo	DC generator	Silencer	Muffler
Earth	Ground	Spanner	Wrench
End float	End play	Split pin	Cotter pin
Engineer's blue	Machinist's dye	Stanchion	Tube
Exhaust pipe	Header	Sulphuric	Sulfuric
Fault diagnosis	Trouble shooting	Sump	Oil pan
Float chamber	Float bowl	Swinging arm	Swingarm
Footrest	Footpeg	Tab washer	Lock washer
Fuel/petrol tap	Petcock	Top box	Trunk
Gaiter	Boot	Torch	Flashlight
Gearbox	Transmission	Two/four stroke	Two/four cycle
Gearchange	Shift	Tyre	Tire
Gudgeon pin	Wrist/piston pin	Valve collar	Valve retainer
Indicator	Turn signal	Valve collets	Valve cotters
Inlet	Intake	Vice	Vise
Input shaft or mainshaft	Mainshaft	Wheel spindle	Axle
Kickstart	Kickstarter	White spirit	Stoddard solvent
Lower leg	Slider	Windscreen	Windshield
Mudguard	Fender		

Chapter 1 Engine

Contents

Specifications

Engine

Type	Twin cylinder four stroke, with pushrod operated overhead valves
Cylinder head	Aluminium alloy (RR 53B)
Valve seat angles	45° inlet and exhaust
Cylinder barrel	Cast iron
Bore	
750 cc	73 mm (2.875 in)
850 cc	77 mm (3.04 in)
Stroke	
750 cc and 850 cc	89 mm (3.503 in)
Capacity (actual)	
750 cc	745 cc
850 cc	828 cc
Compression ratio	
Standard 750 cc	9 : 1
750 cc Combat	10 : 1
850 cc	8.5 : 1
Bhp	
Standard 750 cc	60 @ 6800 rpm
750 cc Combat engine	65 @ 6500 rpm
850 cc	60 @ 5900 rpm

Pistons
Diameter at bottom of skirt ... ...   2.8713 - 2.8703 in (72.931 - 72.906 mm) - 750 cc

Piston rings
Number ... ...   Two compression, one oil control per piston
End gaps ... ...   0.010 - 0.012 in (0.245 - 0.305 mm) - top ring (chrome)
                                                             0.008 - 0.012 in (0.203 - 0.305 mm) - 2nd ring (taper)

Valves
Material ... ...   EN52 - inlet
                                                             KE965 - exhaust
Head diameter ... ...   1.500 in (38.1 mm) inlet - 750 cc
                                                             1.312 in (33.325 mm) exhaust - 750 cc
Stem diameter ... ...   0.3115 - 0.3105 in (7.912 - 7.886 mm) inlet and exhaust valves - 750 cc

Valve springs
Number ... ...   Two per valve
Free length ... ...   1.618 in (41.097 mm) - outer spring
                                                             1.482 in (37.642 mm) - inner spring

Valve guides
Internal diameter ... ...   0.3145 - 0.3135 in (7.988 - 7.962 mm)
External diameter ... ...   0.5015 - 0.50 in (12.738 - 12.725 mm)

Heat resisting washer
Material ... ...   Tufnol ASP
Thickness ... ...   0.062 in (1.574 mm)

Valve timing (measured at 0.013 in (0.3302 mm) cam lift)

	750 cc engines	
	Standard	Combat
Inlet opens BTDC	50°	59°
Inlet closes ABDC	74°	89°
Exhaust opens BBDC	82°	88°
Exhaust closes ATDC	42°	60°

Ignition timing ... ...   28° fully advanced

Main bearings
Make ... ...   FAG   Ransome & Marles
Type NJ306E   6MRJA30
                                                             Existing bearings must be replaced with either of these types
Drive side ... ...   Single lipped roller (pre-1972)
                                                             Special roller (single lipped) 1972 onwards
Timing side ... ...   Single row ball (pre-1972)
                                                             Special roller (single lipped) 1972 onwards
Size ... ...   30 mm x 72 mm x 19 mm (all)

Cylinder head gaskets (750 cc only)

Norton Villiers part number	063844	064071	064072
Type	Eyeletted	Copper	Alloy
Thickness (inches)	0.030	0.040	0.080
Use	750 cc	750	750
	High octane fuel	Reduced compression ratio	Discontinued used for low octane fuel only

850 cc models use Part No 065051

Torque wrench settings	ft lb	Kg m
Cylinder head nuts and bolts (3/8 in)	30	4.15
Cylinder head bolts (5/16 in)	20	2.75
Cylinder base nuts (3/8 in)	25	3.45
and cylinder through bolts (850 cc only)		
Cylinder base nuts (5/16 in)	20	2.75
Connecting rod nuts	25	3.45
Crankshaft nuts	25	3.45
Engine mounting bolts	25	3.45
Rotor nut (alternator)	80	11.06
Alternator stud nuts...	15	2.07
Clutch centre nut	70	9.68
Gearbox sprocket nut	80	11.06
Rear suspension unit mounting nuts	30	4.15
Disc brake caliper fork leg mounting bolts	30	4.15
Disc brake caliper end plug	26	3.60
Oil pressure release valve	25	3.45

1 General description

The engine fitted to the Norton 750 cc and 850 cc Commando models is a vertical twin cylinder, in which both pistons are arranged to rise and fall in unison. The engine is mounted in an inclined position, using what is known as the Isolastic suspension system to damp out the unwanted effects of high frequency engine vibration associated with this type of engine design. An overhead valve layout is used, which is actuated by pushrods that pass through cast-in tunnels at the front of the cylinder barrel.

Lubrication is effected on the dry sump principle, in which oil is fed by gravity to a gear-type pump and distributed to the various parts of the engine. A separate scavenge pump which forms part of the oil pump assembly ensures oil which drains back into the crankcase is returned to the oil tank. The crankshaft assembly comprises a wide centre flywheel, with two outer bob weights one on each side. The big end bearings are of the shell type, fitted to split, light alloy connecting rods. The cylinder barrel is of cast iron and the cylinder head of aluminium alloy, both taking the form of monobloc castings. The Combat engine is normally identified by the additional matt black finish given to the exterior cooling surfaces of the cylinder barrel.

Ignition is provided by twin coils and a twin contact breaker assembly driven off the end of the camshaft, the latter incorporating an automatic advance and retard nut. Twin carburettors are specified, fitted with a large capacity air cleaner. The engine has twin exhaust pipes and silencers of the downswept type with special rubber mountings. Only the 'S' model has an upswept system on the left hand side of the machine.

The lubrication system is protected by a gauze filter within the oil tank and a magnetic drain plug in the crankcase and a pressure relief valve. Post-1972 models have an additional cartridge-type oil filter, with a renewable cartridge element, fitted within the rear engine plates, in close proximity to the rear wheel. Provided the machine is maintained regularly in an intelligent manner, the lubrication system is unlikely to give trouble during normal service.

2 Operations with engine in frame

1 It is not necessary to remove the engine from the frame unless the crankshaft assembly, the big ends or the camshaft require attention. Most operations can be accomplished with the engine in the frame, such as:

a) Removal and replacement of the cylinder head.
b) Removal and replacement of the cylinder barrel and pistons.
c) Removal and replacement of the clutch and primary drive.
d) Removal of the timing sprockets and oil pump.
e) Removal of the contact breaker and automatic advance assembly.

2 When several operations have to be undertaken simultaneously, such as during an extensive rebuild or overhaul, it is often advantageous to remove the engine from the frame after some preliminary dismantling. This will give the advantage of better access and more working space, especially if the engine is attached to a bench-mounted stand.

3 Operations with the engine removed

1 Removal and replacement of the main bearings.
2 Removal and replacement of the crankshaft assembly.
3 Removal and replacement of the camshaft.

4 Method of engine removal

1 The engine is heavy, even for two mechanics. In consequence, it is preferable to partially strip the engine before it is released from the frame, in order to shed some of the heavier components such as the cast iron cylinder barrel and to give greater freedom of movement.
2 The engine and gearbox are separate units which means that the engine can be removed without disturbing the gearbox apart from the need to dismantle the primary transmission.

5 Dismantling the engine - removing the petrol tank

1 Place the machine on the centre stand and ensure that it is standing firmly on level ground.
2 Turn off both petrol taps and disconnect the petrol pipes by unscrewing the unions where they join the base of each tap.
3 Remove the dual seat. This is released by unscrewing the two large diameter milled knobs at the top of each rear suspension unit. The seat will lift straight up in the case of the Fastback models; on other models it is necessary to lift upwards, then pull towards the rear of the machine.
4 To obviate the risk of fire, remove the fuse from its holder in the battery negative lead. This will isolate the electrical system.
5 The petrol tank is secured at the front by self-locking nuts attached to studs which protrude from the underside of the tank. Remove the nuts and any rubber washers below the tank mounting lugs.
5 The rear of the tank will be secured by either a rubber strap or a cross strap beneath the main tube of the frame which is held by nuts and washers. The method of mounting used will depend on the model. When the front and rear fixings have been removed, the tank can be lifted away from the frame.
6 Since there is no necessity to drain the tank prior to these operations, it can represent a dangerous fire hazard. Make sure it is placed well away from the machine and any naked flames or other sources where involuntary ignition may occur.
7 Note the position of the rubber support pads over the main frame tube so that they are located accurately when the tank is eventually replaced.

6 Dismantling the engine - removing the cylinder head

1 Most models are fitted with a downswept exhaust system. This is removed by bending back the tab washer on each exhaust locking ring and then unscrewing each ring to free the exhaust pipe from the cylinder head. Norton Villiers service tool 063968 is recommended for this purpose as the rings are locked tight. If the service tool is not available, the rings can be slackened by careful application of a flat nosed punch and hammer. The 850 cc models have a balance pipe, the clips of which must be slackened.
2 To release the silencers, unscrew the nuts which secure the mounting plates to the two rubber mountings. The exhaust pipes and silencers can then be lifted away as a complete unit.
3 A somewhat similar procedure is recommended for the upswept exhaust system fitted to the 'S' models. In this instance, the silencer has only one point of attachment, on a bracket adjacent to the rear suspension unit.
4 Note there is a copper/asbestos sealing ring in each exhaust port, which should be removed and discarded. It is customary to fit new replacements when the exhaust system is eventually refitted in order to preserve a leaktight joint.
5 Remove both carburettors complete with spacers by unscrewing the four socket screws which retain the assembly to the cylinder head. Access is made easier by using a short socket key in order to provide clearance with the frame tube. Disengage the air cleaner hoses from each carburettor intake and lift the carburettors away. If desired, the carburettors can be separated by disconnecting the balance pipe which joins them and by removing each carburettor top, retained by two crosshead screws. This will enable the slide and needle assembly, complete with control cables, to be lifted out of each mixing chamber.
6 Whichever method is used for the removal of the carburettor,

5.2 Disconnect petrol pipes at petrol tap unions

6.1 Unscrew exhaust locking rings to release downpipes

6.2 Swinging arm prevents removal of mounting bolt ...

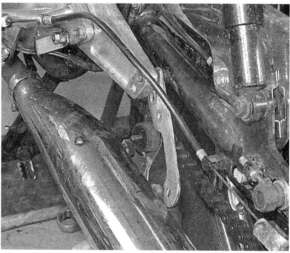
6.2a ... disconnect at rubber mountings as alternative

6.5 Remove carburettors complete with spacers

6.6 Separate air cleaner after removing two retaining bolts

6.7 Twin coil assembly bolts to mounting on frame tube ...

6.7a ... place out of harms way; no necessity to detach completely

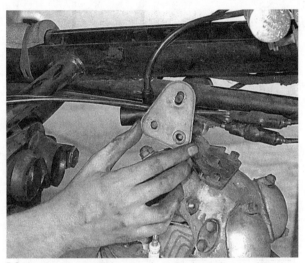

6.8 Head steady assembly comprises two short plates and ...

6.8a ... base plate attached to head by socket screws

6.12 Push rods must be fed into cylinder head to aid removal

7.1 Support pistons as they emerge from bores to prevent damage

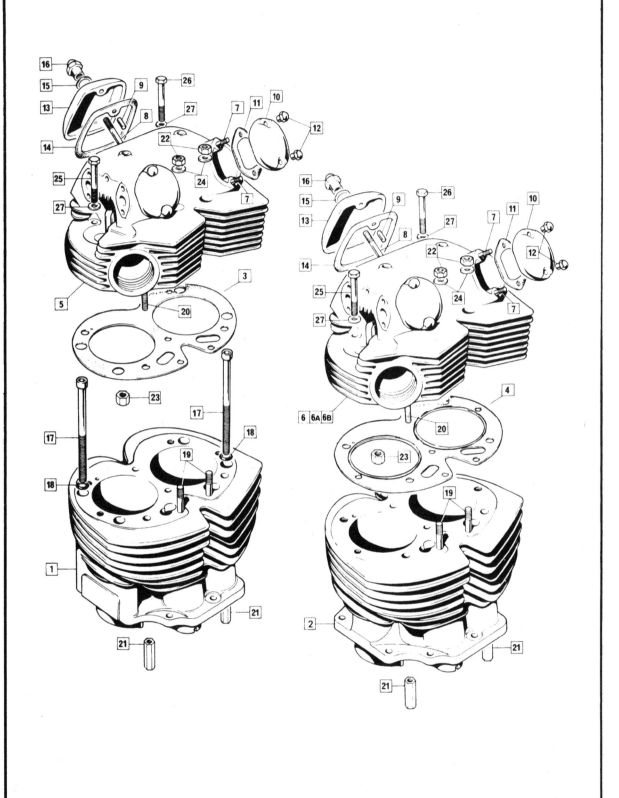

FIG. 1.1. CYLINDER BARREL AND CYLINDER HEAD

1 Cylinder barrel - 850 cc models
2 Cylinder barrel - 750 cc models
3 Cylinder head gasket (eyeletted) - 850 cc models
4 Cylinder head gasket (eyeletted) - 750 cc models
5 Cylinder head - 850 cc models
6 Cylinder head - 750 cc models (30 mm, low compression - marked RH1)
6a Cylinder head - 750 cc models (32 mm, high compression - marked RH6)
6b Cylinder head - 750 cc models (32 mm, low compression - marked RH5)
7 Rocker cover stud - front
8 Rocker cover stud - rear
9 Rocker cover dowel
10 Rocker cover - front 2 off
11 Rocker cover gasket - front 2 off
12 Rocker cover nut - front 4 off
13 Rocker cover - rear
14 Rocker cover gasket - rear
15 Rocker cover stud washer
16 Rocker cover stud nut - rear
17 Cylinder through bolt (850 cc models only) 4 off
18 Washer for through bolt (850 cc models only) 4 off
19 Cylinder barrel stud 2 off
20 Cylinder head stud 3 off
21 Cylinder head sleeve nut 2 off
22 Cylinder head nut 2 off
23 Cylinder head nut
24 Cylinder head washer 2 off
25 Cylinder head bolt 4 off
26 Cylinder head bolt (short)
27 Cylinder head washer 5 off

ensure the parts involved are taped out of harms way. They are very easily damaged if mishandled. Lift away the air cleaner assembly by withdrawing the two long bolts which hold the case together.

7 Unbolt the twin ignition coil assembly from its mounting above the cylinder head and disconnect both caps from the spark plugs. The coil assembly need not be removed completely; it can be tied to the handlebars so that it does not impede removal of the cylinder head. Only the electrical connections from the coils to the contact breakers need be removed.

8 Remove the cylinder head steady. This takes the form of two short plates which interconnect a plate bolted to the cylinder head with two rubber mountings, one on each side of the tube below the main frame tube. Remove the nuts from the rubber mountings first in order to prevent the mountings from rotating. When the two mounting plates have been removed, remove also the head steady from the cylinder head by withdrawing the three socket screws which retain it.

9 Remove both spark plugs and remove the rocker oil feed pipes from both sides of the cylinder head casting. Special care is needed when unscrewing the banjo union bolts to prevent the thin copper pipe from twisting or necking. Detach the copper sealing washers from either side of each union but leave the feed pipe attached to the timing cover.

10 Slacken and remove nine of the cylinder head retaining bolts, leaving only the front centre bolt in position. This and the two bolts, one on either side, are recessed into the cylinder head and will require a slim socket or box spanner for their release. Two nuts are found on the underside of the cylinder head, at the front, and another single nut on the underside, at the rear. The four remaining bolts are easily accessible from the top of the cylinder head, making a total of ten bolts. 850 cc models include four through bolts in the cylinder barrel.

11 When the front cylinder head bolt is removed last of all, the cylinder head will tilt a little against the spring pressure of the valve which is open. This will aid breaking the cylinder head joint.

12 Before the cylinder head can be lifted away, it is necessary to feed each of the four pushrods into the cylinder head as far as possible, after detaching them from the ends of the rocker arms. This can be accomplished by tilting the cylinder head towards the rear, whilst holding the cylinder head with one hand and the pushrods with the other. Do not use force and make sure the pushrods are clear of the cylinder barrel as the head is being removed. Failure to observe this precaution may cause damage to the light alloy pushrods, necessitating their renewal.

13 The cylinder head gasket will adhere to either the cylinder barrel or cylinder head and should not be re-used unless it is completely undamaged.

7 Dismantling the engine - removing the cylinder barrel and pistons

1 The cylinder barrel is retained to the crankcase by nine studs (and four through bolts, 850 cc models only). Remove the nine holding down nuts (and the through bolts, 850 cc models) and ease the cylinder barrel upwards, taking care to support the pistons as they emerge from the cylinder bores. If broken piston rings or other piston damage is suspected, it is advisable to pad the mouth of the crankcase with clean rag immediately the cylinder barrel is raised to prevent displaced particles from falling in.

2 Remove and discard the circlips from each piston, then remove both gudgeon pins, taking care to support the piston and connecting rod as they are tapped out of position. If the pins are a tight fit, the pistons should be warmed first by placing a rag soaked in hot water on each crown. This will expand the alloy of the piston and release the grip of the gudgeon pin boss. On no account use force, or the light alloy connecting rods may be damaged permanently.

3 Mark each piston INSIDE the skirt to ensure it is replaced in its original position. The pistons are individually marked on the crown to this effect, but the marks may have been erased if the

crown was badly scratched during a previous decoke. Although the valve cutaways are the same size in the crown of the standard ratio pistons, the cutaway for the exhaust valve is positioned much nearer to the edge. This makes the correct location of the pistons vital.

8 Dismantling the engine - removing the alternator, clutch and primary chaincase

1 Remove the three nuts and washers which secure the left hand footrest and rear brake pedal to the light alloy mounting plate. In order to clear the primary chaincase cover, the footrest assembly can be left hanging from the brake cable; the alternative is to remove the electrical snap connector from the stop lamp switch on the brake pedal and the brake cable from the rear brake operating arm so that the complete assembly can be detached and lifted away.

2 Place a large tray immediately below the underside of the chaincase joint. The chaincase has no drain plug and in consequence the full oil content will be released immediately the two halves are separated.

3 Remove the centre sleeve nut which retains the two chaincase halves together and rock the chaincase slightly to clear the two locating dowels, one at the top and one at the bottom. Immediately the seal breaks, the oil content will be released into the tray. Remove the outer cover completely, taking care that the rubber seal around the jointing face or rear half of the chaincase is not displaced and damaged.

4 Place a stout metal rod through the small end of both connecting rods and rotate the engine in an anticlockwise (drive) position until the rod rests across the crankcase mouth. This will lock the engine in position so that the nut of the alternator rotor can be slackened and removed. Use a ½ inch Whitworth socket wrench and turn in an anticlockwise direction; the nut has a normal right hand thread. Do not misplace the washer which seats below the nut.

5 Remove the alternator stator which is secured by three nuts and washers. Draw the stator assembly off the studs, after detaching the lead wire at the snap connectors in the vicinity of the air cleaner housing. The lead wire will pull through the small rubber grommet in the centre of the rear chaincase.

6 The alternator rotor is keyed onto the crankshaft and has a parallel fit. In consequence, it is not difficult to remove. Light pressure with a pair of tyre levers positioned at the rear of the rotor should provide sufficient leverage. Remove the rotor key, the packing collar and any shims on the crankshaft. Detach the three spacers from the stator mounting studs and place them in a safe position for reassembly.

7 Whilst the engine is still locked in position, slacken the clutch pushrod adjuster nut in the centre of the clutch assembly and remove both the adjuster and the nut.

8 The clutch cannot be dismantled without the compressor which is necessary for the diaphragm spring. Norton Villiers service tool 060999 is recommended for this purpose. DO NOT ATTEMPT TO DISMANTLE THE CLUTCH WITHOUT A COMPRESSOR. IF THE TENSION OF THE DIAPHRAGM SPRING IS RELEASED SUDDENLY WITHOUT PROTECTION, SERIOUS INJURY MAY RESULT. Do not unscrew the compressor from the diaphragm or the latter will be released with considerable force.

9 Screw the centre bolt of the compressor into the hole previously occupied by the clutch pushrod adjuster and check that at least ¼ inch of the bolt has engaged with the internal thread. Turn the nut in a clockwise direction until the diaphragm spring is flat and free to rotate. This will enable the retaining circlip to be prised from its groove inside the periphery of the clutch body by a screwdriver blade. Lift the first end clear of the groove, then peel the circlip out of position. The compressor and diaphragm can then be lifted away together; there is no necessity to detach the compressor until after the clutch is reassembled. NEVER peel out the circlip without first using a compressor to compress the diaphragm.

10 Lift out the clutch plates, using two pieces of stout wire

7.2 Never re-use circlips. Old circlips must be discarded

7.2a Suport the piston and withdraw the gudgeon pin

8.5 Alternator stator is secured by three nuts

8.5a Lead wire will pull through grommet in chaincase

8.6 Alternator rotor is keyed on to crankshaft

8.7 Slacken push rod adjuster whilst engine is locked

with their ends bent at a right angle, to form a hook. There is a total of eight, four plain and four friction. The clutch inner drum can now be released by unscrewing the centre nut. To prevent the clutch from turning, select top gear and apply the rear brake by pressing on the operating arm. The centre nut has a right hand thread and should be removed, after bending back the tab washer, complete with the spring washer beneath it. Pull out the clutch pushrod. The clutch inner drum can now be withdrawn.

11 The triplex primary chain is of the endless type and has no split link joint. In consequence it is necessary to remove the clutch outer drum with its integral sprocket and the engine sprocket in unison. The engine sprocket is a keyed taper fit on the end of the crankshaft and it is essential to use a sprocket puller to achieve its release. Norton Villiers service tool 060941 is specified for this purpose; the sprocket is tapped to accept the extractor bolts. If the service tool is not available, a two or three-legged sprocket puller can be used with equal effect. Lift away both sprockets together with the triplex chain. Take special care of the collar and spacers fitted over the gearbox mainshaft, behind the clutch, since they determine the accurate alignment of the two sprockets. Place them in a safe place until reassembly commences. There are spacers on the crankshaft, behind the engine sprocket location, and a Woodruff key.

12 The rear half of the chaincase is attached to the left hand crankcase by three screws secured by tab washers. Bend back the tabs, remove the screws, and pull away the rear chaincase. Remove the spacers used on the centre stud which engages with the sleeve nut that retains the chaincase. The overall length of the spacers must be correct to prevent distortion of the chaincase when the sleeve nut is tightened.

9 Dismantling the engine - removing the crankcase assembly from the frame

1 Unscrew the tachometer cable from the union joint in front of the right hand cylinder barrel. At the same time it is convenient to drain the oil from the oil tank by removing the drain plug, or in the case of the early models, by taking out the larger oil filter union which secures the main oil feed pipe from the oil tank. It will be necessary to detach the right hand side cover in order to gain access to the oil tank. This task is best accomplished whilst the oil is warm, so that it will flow more freely.

2 Remove the rocker feed pipe which is joined to the timing cover by a banjo union. Take care not to lose the copper sealing washers.

3 Remove the gear indicator from the gearbox end cover which is retained by a setscrew and the gear change lever, secured on splines by a pinch bolt. Push the crankcase breather pipe clear of the oil tank. Remove the right hand footrest secured by two nuts and a bolt.

4 Place a large capacity tray under the crankcase assembly and remove the oil pipe junction block from the right hand crankcase, assuming the oil tank has drained completely. Slacken and remove the hexagon headed drain plug in the bottom of the left hand crankcase. Early models (engine numbers prior to 200000 and those of 850 cc capacity) have a large diameter plug (7/8 inch Whitworth) which houses a separate filter. Later 750 cc models have a much smaller plug of the magnetic type which obviates the need for a filter. This is an additional fitting on the 850 cc engines. Allow all excess oil to drain off. Take off the crankcase breather (early models only).

5 It is necessary to remove the crankcase assembly by detaching the front engine mounting. Remove the large diameter bolt which passes through the centre of the mounting, taking care to align the flats on the head so that the bolt will clear the timing case during removal. The bolt has a small head to facilitate its removal in this fashion. Prise back the gaiter on one end of the engine mounting so that the spacer, end cap and shims can be removed to ensure freedom of movement within the frame lugs.

6 Remove the two nuts from the right hand side of the front engine mounting studs, then pull out the studs complete with the remaining nuts from the left hand side of the mounting. The

8.10 Release inner drum by unscrewing retaining nut

8.11 Puller is necessary to free engine sprocket

8.11a Chain is endless. Take off sprockets together

front engine mounting can now be withdrawn from the crankcase assembly and from the frame.

7 Remove the two lower rear engine bolts which pass through the engine plates encasing the gearbox. Before the lower of the two bolts can be withdrawn, it will be necessary to lift the front of the crankcase assembly upwards so that the bolt will clear the lower frame tubes. Note the earthing connection to the frame. It is advisable to keep the drain tray below the engine during this operation since if the drain plug is removed, some additional oil may escape from the crankcase as the engine position is changed.

8 Support the crankcase assembly and withdraw the upper rear engine bolt. The crankcase assembly is now free to be removed from the frame and can be lifted out from the right hand side of the machine, leaving the gearbox in position.

10 Dismantling the engine - removing the contact breaker and auto-advance assembly

1 With the crankcase assembly out of the frame, it is preferable to devise some form of bench-mounting engine stand so that the engine can be held rigidly on the workbench during the final stages of dismantling and during the early stages of reassembly. This will have the advantages of leaving both hands free to continue the dismantling work and providing better access. As an alternative, remove the long bolt which passes through the base of the crankcase assembly and grip the unit in a vice fitted with soft aluminium clamps.

2 Remove the two screws which retain the circular contact breaker cover in position and lift away the cover. Remove the centre bolt of the contact breaker cam, the serrated washer and the plain washer. The cam can now be extracted from its taper by using Norton Villiers service tool 060934. If this tool is not available, the cam can be released by screwing a 5/16 inch UNF bolt into the thread tapped within the cam orifice.

3 There is no necessity to remove the contact breaker baseplate before the timing cover is removed but if attention to this component is necessary, it is secured by either two screws or two hexagonal pillars. Unscrew the two screws or pillars and lift the contact breaker assembly away. The contact breaker lead wire will pull through the passageway in the contact breaker housing.

11 Dismantling the engine - removing the timing cover, oil pump and timing chain

1 Remove the twelve screws which retain the timing cover in position and free the cover by giving a gentle tap from behind the pressure release valve with a rawhide mallet. Lift the timing cover away and withdraw the two locating dowels at the front and rear of the timing chest which may otherwise be displaced during the following operations.

2 Remove the two nuts which retain the oil pump. The pump may be a tight fit on the two mounting studs and is best released by rotating the oil pump drive pinion so that it will travel along the worm drive of the crankshaft and ease itself off the studs.

3 The drive worm is integral with the left hand thread nut on the end of the crankshaft. Lock the engine by placing a stout metal bar through the small ends of both connecting rods, then slacken the nut in a CLOCKWISE direction with a socket spanner. The nut is tight and some force may be necessary to start it moving.

4 Whilst the engine is still locked in position, slacken the nut retaining the camshaft sprocket. This has a normal right hand thread and it will be necessary to re-arrange the method of locking the engine since this nut will turn in the opposite (anti-clockwise) direction. Use steady and not excessive force, otherwise the intermediate gear spindle may be forced out of the crankcase.

5 Because the timing chain is of the endless variety, the two sprockets and chain must be removed in unison. If the camshaft sprocket is a little tight on the end of the camshaft, it can be

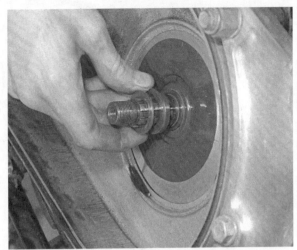

8.11b Note spacers behind clutch for correct chain alignment

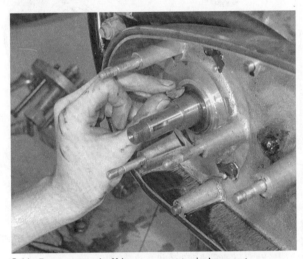

8.11c Remove woodruff key to prevent misplacement

8.12 Chaincase rear is retained by three bolts and tab washers

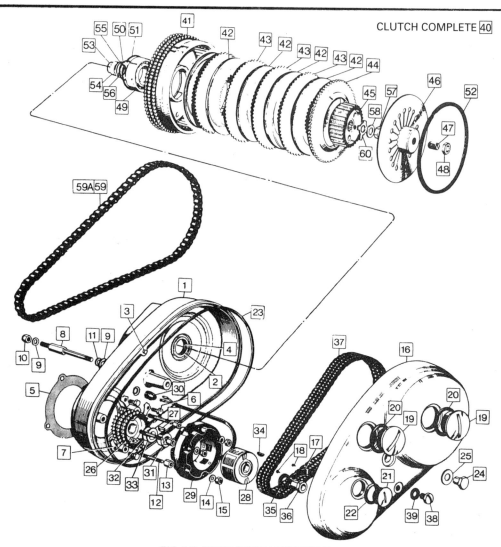

FIG. 1.2. PRIMARY TRANSMISSION

1 Inner chaincase complete with dowels
2 Chaincase oil seal disc 2 off
3 Chaincase dowel 2 off
4 Chaincase felt seal
5 Chaincase gasket
6 Chaincase bolt 3 off
7 Tab washer 3 off
8 Chaincase centre stud
9 Chaincase centre stud washer 2 off
10 Chaincase centre stud nut
11 Chaincase centre stud shim - number as required
12 Stator stud 3 off
13 Stator stud spacer 3 off
14 Stator stud washer 3 off
15 Stator stud nut 3 off
16 Outer chaincase
17 Ignition indicator plate
18 Hammer drive screw 2 off
19 Chain inspection cap
20 'O' ring for chain inspection cap
21 Timing inspection cap
22 'O' ring for timing inspection cap
23 Chaincase sealing rubber
24 Chaincase attachment nut
25 Chaincase attachment nut washer
26 Engine sprocket
27 Engine sprocket key
28 Alternator rotor
29 Alternator stator
30 Alternator leadwire grommet
31 Rotor spacer

32 Rotor shim 0.010 in. number as required
33 Rotor shim 0.036 in. number as required
34 Rotor key
35 Rotor washer
36 Rotor nut
37 Primary drive chain
38 Chaincase oil level plug
39 Oil level plug 'O' ring
40 Clutch assembly complete
41 Clutch sprocket complete with backplate
42 Clutch friction plate 4 or 5 off, depending on type
43 Clutch plain plate 3 or 4 off, depending on type of
    friction plate
44 Clutch pressure plate
45 Clutch centre
46 Clutch diaphragm with centre
47 Clutch adjuster
48 Clutch adjuster locknut
49 Clutch centre bearing
50 Clutch bearing inner circlip
51 Clutch bearing outer circlip
52 Clutch diaphragm circlip
53 Clutch location circlip
54 Clutch location spacer
55 Clutch location shim 0.036 in. number as required
56 Clutch location shim 0.048 in. number as required
57 Clutch retaining nut
58 Clutch retaining nut washer
59 Final drive chain - 98 links, 19 tooth sprocket or 99
    links, 21 tooth sprocket
60 Clutch retaining nut tab washer

eased off with a pair of tyre levers, taking care not to bruise the sealing face of the timing cover joint.

12 Dismantling the engine - removing the crankshaft pinion and separating the crankcases

1 It is essential to use Norton Villiers service tool ET 2003 to remove the crankshaft pinion because it is necessary to have a straight pull. There is insufficient room behind the pinion to insert a conventional two or three legged puller. Any other method of removal will almost inevitably cause damage.

2 When the pinion has been withdrawn, the Woodruff key and backing washer can be lifted off the crankshaft. There is a lipped oil sealing disc which tends to cling to the right hand main bearing due to the residual oil film. This washer is best removed by two small magnets. If desired, the chain tensioner can be removed at the same time. It is held by two bolts and washers. Mark its position to ensure correct chain tension on replacement.

3 The crankcase assembly must now be removed from either the engine stand or vice in which it is held, so that the crankcases can be parted. They are held together by two studs, one bolt and two setscrews. When all of these fixings have been removed, part the crankcases by tapping with a bar of wood (such as a hammer handle) against the left hand crankcase mouth. It will be noted that as the left hand crankcase is lifted away, the rotary breather disc and spring will be displaced from the camshaft bush (pre-1972 models).

4 Withdraw the camshaft from the right hand crankcase, taking care not to lose the chamfered thrust washer. Engines with a number above 200000 have an arrangement comprising two thrust washers with tabs that engage with holes in the crankcase.

5 It is preferable to ease the right hand crankcase from the crankshaft rather than vice versa. The most effective means, which obviates striking the end of the crankshaft, takes the form of a spacer in the form of a hollow tube which abuts against the crankshaft shoulder. If the crankcase is rested on the workbench, inner face uppermost, the spacer will permit the crankcase to be driven off the crankshaft by hammering downwards on a block of soft wood pressed against the crankcase casting. This is an operation requiring the assistance of a second person, one to hold the crankcase and crankshaft assembly and one to drive the crankcase in a downwards direction. The alternative is to heat the crankcase so that the right hand main bearing is displaced with the crankshaft assembly, to be drawn off at a later stage, if desired.

6 Engines with a number over 200000 are fitted with a roller bearing in the right hand crankcase and in this instance there is no problem in removing the crankshaft assembly. This also applies to the 850 cc.

7 The camshaft bushes need not be removed unless wear necessitates their renewal. The bushes have an extremely long life under normal service and require special machining facilities when renewal is necessary, a task requiring expert attention.

8 The main bearings are best removed from the crankcases by heating the area around the bearing housing and either drifting them out of position or by bumping the jointing face against a flat wooden surface so that they are displaced by the shock. A blow lamp is particularly suitable for this task, but on no account use a welding torch otherwise the alloy will melt.

13 Examination and renovation - general

1 Before examining the parts of the dismantled engine for wear, it is essential that they should be cleaned thoroughly. Use a petrol/paraffin mix to remove all traces of old oil and sludge that may have accumulated within the engine and a cleansing agent such as Gunk or Jizer for the external surfaces. Special care should be taken when using these latter compounds, which require a water wash after they have had time to penetrate the film of grease and oil. Water must not be allowed to enter any of the internal oilways or parts of the electrical system.

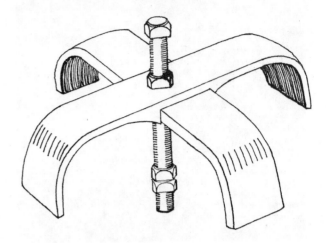

Fig. 1.3. Tool for compressing clutch diaphragm spring

9.2 Rocker feed pipe is attached to rear of timing cover

9.3 Rear bolt acts as earthing connection, behind alloy plate

9.4 Early models have breather attached to crankcase

9.5 Align head of mounting bolt to clear crankcase casting

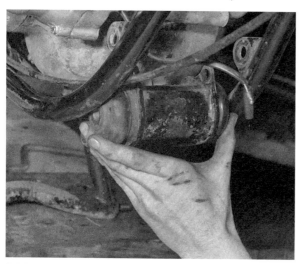

9.6 Pull engine mounting downwards to clear frame

9.7 Note earth connection to lower rear engine bolt

9.8 Tilt engine forwards, then ...

9.8a ... lift out of engine plates

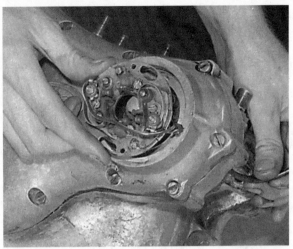

10.2 Remove two retaining screws to free base plate assembly

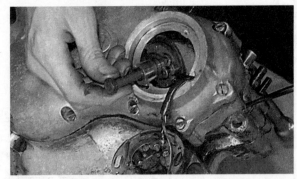

10.2a Use a 5/16 in UNF bolt to extract cam and auto-advance unit

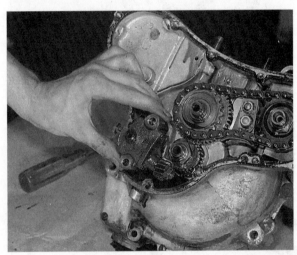

11.2 Oil pump is retained by two nuts

11.3 Crankshaft pinion has a LEFT HAND thread

2   Examine the crankcase castings for cracks or other signs of damage. If a crack is discovered, it will require specialist repair, or the renewal of both crankcases. Crankcases are supplied in matched pairs since it is considered bad engineering practice to renew only one. Under these latter circumstances there can be no guarantee that the main bearing housings have been bored exactly in line with one another.
3   Examine carefully each part to determine the extent of wear, if necessary checking with the tolerance figures listed in the Specifications section of this Chapter. The following sections of this Chapter describe how to examine the various engine components for wear and how to decide whether renewal is necessary.
4   Always use a clean, lint-free rag for cleaning and drying the various components prior to reassembly, otherwise there is risk of small particles obstructing the internal oilways.

14 Crankshaft, big end and engine bearings - examination and renovation

1   Check the big end bearings for wear by pulling and pushing on each connecting rod in turn, whilst holding the rod under test in the vertical plane. Although a small amount of side play is permissible, there should be no play whatsoever in the vertical direction if the bearing concerned is fit for further service.
2   The big end bearings take the form of shells. To gain access to the shells, remove the connecting rods by unscrewing the two self-locking nuts at each end cap. When the nuts have been withdrawn completely, the connecting rod and end cap can be pulled off the crankshaft, with the bearing shells still attached. Mark both the connecting rods and their end caps clearly so that there is no possibility of them being interchanged. Note that the locating tabs of the bearing shells fit to the same side of each connecting rod.
3   If the crankshaft assembly is to be separated, it is advisable to continue the dismantling in a clean metal tray. The assembly

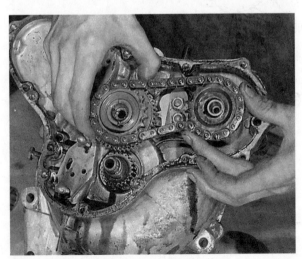

11.5 Lift off timing chain and sprockets in unison

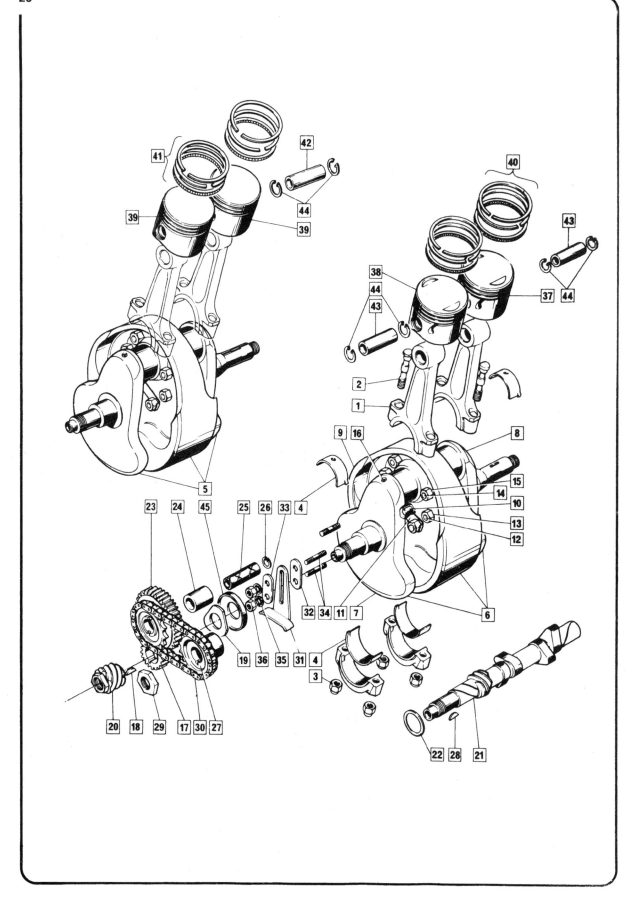

FIG. 1.4. CRANKSHAFT ASSEMBLY AND TIMING GEAR

1 Connecting rod with cap 2 off	24 Intermediate gear bush
2 Connecting rod bolt 4 off	25 Intermediate gear shaft
3 Connecting rod self-locking nut 4 off	26 Intermediate gear shaft circlip
4 Big-end shell set 2 off	27 Camshaft sprocket
5 Crankshaft complete - 850 cc models	28 Camshaft sprocket key
6 Crankshaft complete - 750 cc models	29 Camshaft sprocket nut
7 Crank cheek, timing side - 750 cc models only	30 Camshaft chain
8 Crank cheek, drive side - 750 cc models only	31 Tensioner slipper
9 Flywheel - 750 cc models only	32 Tensioner backing plate - thin
10 Dowel	33 Tensioner backing plate - thick
11 Nut retaining plate 2 off	34 Tensioner stud 2 off
12 Crankshaft stud 4 off	35 Tensioner stud washer 2 off
13 Crankshaft stud nut 8 off	36 Tensioner stud nut 2 off
14 Crankshaft stud 2 off	37 Piston complete (left hand) 750 cc only
15 Crankshaft stud nut 4 off	38 Piston complete (right hand) 750 cc only
16 Crankshaft oilway screw	39 Piston complete - 850 cc models only 2 off
17 Crankshaft pinion	40 Piston ring set - 750 cc models only 2 off
18 Crankshaft pinion key	41 Piston ring set - 850 cc models only 2 off
19 Crankshaft pinion backplate	42 Gudgeon pin - 850 cc models only 2 off
20 Oil pump worm	43 Gudgeon pin - 750 cc models only 2 off
21 Camshaft	44 Circlip 4 off
22 Camshaft thrust washers 2 off	45 Lipped sealing washer, timing side bearing
23 Intermediate gear complete with bush and sprocket	

holds approximately one teacup of oil which will be released when the cranks are separated from the centre flywheel.

4 Slacken the nuts on the right hand (timing) side of the crankshaft assembly, noting that two of the nuts are retained by a tab washer and the other four have been centre-punched to prevent them from loosening in service. All will be tight.

5 Mark the central flywheel so that there is no possibility of it being reversed during subsequent reassembly. Jar the crank flanges from the flywheel with a hammer and a soft metal drift. This will release the left hand (drive side) crankshaft, which will come away with the lower two studs, tab washer and nuts. If the machine has covered a considerable mileage, it will probably be found that there is a build up of sludge in both crank flanges and in the flywheel recess as a result of the centrifugal action of the rotating assembly. This must be cleaned out thoroughly prior to reassembly.

6 Wash each crank in turn with clean petrol and use compressed air to dry off. Light score marks on the big end journals can be removed by the use of fine emery cloth but if the scoring is excessive or deep, or if measurements show ovality of more than 0.0015 inch (0.0381 mm) the journals must be reground. The accompanying illustration shows the regrind sizes permissible; shell bearings are available in undersizes to match, from minus 0.010 inch to 0.040 inch, in 0.010 inch stages. The big end shells are finished to give the required diametrical clearance and must not be scraped to improve the fit.

7 Main bearing roller races should be withdrawn from the crankshaft by using Norton Villiers service tool 063970. There is insufficient clearance for the legs of a sprocket puller or other substitute, if the service tool is not available. Main bearing failure is characterised by a rumbling noise from the engine and some vibration which may not be damped out by the Isolastic engine mountings. Bearings of the ball or roller type should be renewed if any play is evident, if the tracks are worn or pitted, or if any roughness is felt when they are rotated by hand.

15 Timing pinions, timing chain and chain tensioner - examination and renovation

1 It is unlikely that either the timing pinions or sprockets will require attention unless a timing chain breakage has resulted in either chipped or broken teeth. These components have an exceptionally long life and rarely need renewing.

2 The timing chain is of the endless type and should be examined closely for any signs of broken rollers or cracked side plates. If the chain has worn unevenly, making it difficult to adjust the chain tensioner, this is another cause for renewal. If there is any doubt about the condition of the chain it is wise to renew it, since a breakage may damage the timing pinions and sprockets, apart from completely immobilising the machine.

3 Although the chain tensioner will be grooved from sliding contact with the side plates of the timing chain, there is no necessity for renewal unless the grooves are deep and there is danger of the rollers making contact.

16 Timing cover oil seals - examination and renewal

1 The inside of the timing cover contains two oil seals, one for the crankshaft and one for the contact breaker assembly. The presence of oil within the contact breaker housing denotes failure of the contact breaker seal which must be renewed. Close examination will show whether the crankshaft seal has remained effective.

2 The crankshaft oil seal is retained in the timing cover housing by a circlip. When the circlip is removed, the seal can be prised from the housing. Fit the new seal with the pressure side towards the timing cover and after driving it home, refit the circlip with its sharp side towards the crankcase. Avoid scratching the oil seal housing during this operation and check that the circlip has located fully with its groove.

3 The contact breaker oil seal is pressed into its housing without any retaining device. Prise the old seal out of position and fit the new seal with the pressure side towards the crankcase (spring side).

4 Since the oil seals are easily damaged if too much force is used when pressing them home, two Norton Villiers service tools are available for this purpose. Use tool 063967 for the crankshaft oil seal and tool 063966 for the contact breaker seal.

17 Cylinder barrel - examination and renovation

1 The usual indications of badly worn cylinder bores and pistons are excessive oil consumption and piston slap, a metallic rattle which occurs when there is little or no load on the engine. If the top of the cylinder barrel is examined carefully, it will be observed that there is a ridge on the thrust side of each cylinder bore which marks the limit of travel of the uppermost piston ring. The depth of this ridge will vary according to the amount of wear that has taken place and can therefore be used as a guide to bore wear.

2 Measure the bore diameter below each ridge, using an internal micrometer. Compare this reading with the diameter at the bottom of each bore. If the difference in readings exceeds 0.005 inch (0.1270 mm) it is necessary to have the cylinder barrel rebored and to fit oversize pistons and rings.

3 If an internal micrometer is not available, the amount of wear can be measured by inserting a compression ring so that it is about ½ inch from the top of the bore and seated squarely in the bore by pressing it down with the skirt of the piston. Measure the ring gap with a feeler gauge, then reposition the ring below the area traversed by the piston and measure the gap again. Subtract the second reading from the first and divide the difference by three to give the diametrical wear. If in excess of 0.005 inch (0.1270 mm) a rebore is necessary.

4 Check the surfaces of both cylinder bores to ensure there are no score marks or other signs of damage that may have resulted from an earlier engine seizure or displacement of one of the circlips. Even if the bore wear is not sufficient to necessitate a rebore, a deep indentation will override this decision in view of the compression leak that will occur.

5 Check that the external cooling fins are not clogged with road dirt or oil, otherwise the engine may overheat. Unlike most other engines fitted with a cast iron cylinder barrel, an aluminium finish is applied in place of the traditional matt black, only the Combat engine has a black-finished cylinder barrel.

6 Examine the base flange of the cylinder barrel. If the engine has been overstressed by excessive tuning, the flange is one of the first parts to fail, usually around the root of the bores. If the flange is cracked, renewal of the cylinder barrel is essential.

7 The cam followers are located in the base of the cylinder block, where each pair are retained by a locating plate and two setscrews, wired together. Unless the hardened surface is badly worn, chipped or has broken through, there is no necessity to remove them. Note, however, that when the camshaft is renewed, the followers must also be renewed as a set, irrespective of their condition.

18 Pistons, piston rings and small ends - examination and renovation

1 If a rebore is necessary, the pistons and rings can be discarded because they must be replaced by their oversize counterparts.

2 Remove all traces of carbon from the piston crowns, using a soft scraper to ensure the surface is not marked. Finish off by polishing the crowns with metal polish, so that the carbon will not adhere so readily. NEVER use emery cloth.

3 Piston wear usually occurs at the base of the skirt and takes the form of vertical streaks or score marks on the thrust side. If a previous engine seizure has occurred, the score marks will be very obvious. Pistons which have been subjected to heavy wear or seizure should be rejected and new ones obtained.

4 The piston ring grooves may become enlarged in use, permitting the rings to have greater side float. It is unusual for

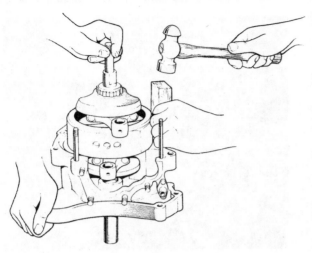

14.4 Slacken nuts on right-hand side of assembly

Fig. 1.5. Supporting the crankcase whilst removing the crankshaft assembly

14.5 Before **crankshaft** assembly is separated, mark flywheel to prevent accidental reversal

16.2 Crankshaft oils seal is driven into position, ...

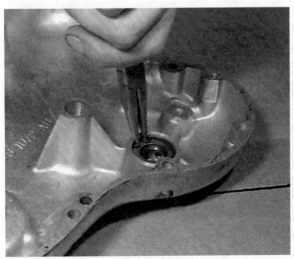

16.2 ... then retained by a circlip

16.3 Fit contact breaker oil seal as shown

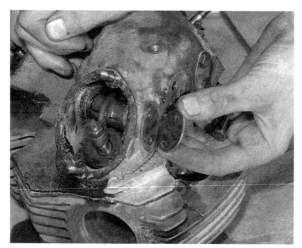

19.1 Remove rocker cover plates first

19.1a Locking plate locates with end of rocker spindle

19.3 Spring compressor can be fitted without removing rocker spindles

this type of wear to occur on its own, but if the side float appears excessive, new pistons of the correct size should be fitted.

5 Piston ring wear is measured as detailed in Section 17.3. If the end gap in the two positions is near identical, but is greater than the recommended limit of 0.012 inch, the piston rings are worn and must be renewed.

6 The gudgeon pins must be a good sliding fit in the small end of the connecting rods without evidence of play. The connecting rods are not bushed and must be renewed if excessive small end wear occurs. Worn small ends produce a rattle, not unlike piston slap, which will rapidly increase in intensity.

19 Valves, valve springs and valve guides - examination and renovation

1 Before the valves, valve springs and valve guides can be examined, it is preferable to remove the rocker arms and valves from the cylinder head to give better access for a valve spring compressor. Commence by withdrawing the rocker spindles, to which access is gained after the cover plates have been removed. Each cover plate is secured by two bolts; it is probable that the cover can be lifted away together with the rocker spindle lock plate and the associated gaskets, as a complete unit. Under these circumstances there is no necessity to separate the individual parts.

2 The rocker spindles are an extremely tight fit in the cylinder

head casting and it will be necessary to use Norton Villiers service tool 061028 to extract them, preferably after the whole cylinder head has been immersed in hot water to help expand the surrounding housing. Note that each rocker arm has a double spring washer against the innermost end of its boss and a plain washer against the outer end. Three thrust washers take up end float and will be displaced, along with the rocker, when the spindle is withdrawn.

3 It is now possible to insert a valve spring compressor and release each of the valves in turn. Keep the valves, valve springs and collets etc together in sets so that they are eventually replaced in their original location. Note that the inlet valve stems have a rubber seal, which must be renewed on reassembly.

4 After cleaning all four valves to remove carbon and burnt oil, examine the heads for signs of pitting or burning. Examine the valve seats in the cylinder head. The exhaust valves and their seats will require the most attention because they are the hotter running. If the pitting is slight, the marks can be removed by grinding the seats and valve heads together, using fine valve grinding compound.

5 Valve grinding is a simple, if somewhat laborious task. Smear a trace of fine valve grinding compound (carborundum paste) on the seat face and apply a suction grinding tool to the head of the valve. Oil the stem of the valve and insert it in the guide until it seats in the grinding compound. Using a semi-rotary motion, grind-in the valve head to its seat, using a backward and forward motion. It is advisable to lift the valve occasionally to distribute the grinding compound more evenly. Repeat this application until an unbroken ring of light grey matt finish is obtained on both valve and seat. This denotes the grinding operation is now complete. Before passing to the next valve, make sure that all traces of the valve grinding compound have been removed from both the valve and its seat and that none has entered the valve guide. If this precaution is not observed, rapid wear will take place due to the highly abrasive nature of the carborundum base.

6 When deep pits are encountered, it will be necessary to use a valve refacing machine and a valve seat cutter, set to an angle of 45°. Never resort to excessive grinding because this will only pocket the valves in the head and lead to reduced engine efficiency. If there is any doubt about the condition of a valve, fit a new one.

7 Examine the condition of the valve collets and the groove on the valve stem in which they seat. If there is any sign of damage, new parts should be fitted. Check that the valve spring collar is not cracked. If the collets work loose or the collar splits whilst the engine is running, a valve could drop in and cause extensive damage.

8 Measure the valve stems for wear, comparing them with the unworn portion that does not extend into the valve guide. Check also the valve guides for excessive play. Valve stem diameter, when new, is 0.3145 - 0.3135 in (7.921 - 7.886 mm). Check that the end of the stem is not indented from contact with the rocker arm, making tappet adjustment difficult.

8 Check the free length of each valve spring and replace the whole set if any has taken a permanent set. The free length is as follows:

Outer springs	1.618 in (41.097 mm)
Inner springs	1.482 in (37.642 mm)

Worn or 'tired' valve springs have a marked effect on engine performance and should preferably be renewed during each decoke as a minimum, especially in view of their low overall cost.
9 The cast iron valve guides are a tight interference fit in the aluminium alloy cylinder head and can be removed and refitted only after the cylinder head has been heated to a temperature in the region of 150° - 200°C. Norton Villiers service tool 063964 is available for removing them, but if the service tool is not available, a double diameter drift to match the valve stem diameter, or a bolt with a rounded head, to form a drift, can be used with equal effect. The cylinder head MUST be heated to the CORRECT TEMPERATURE during this operation.
10 Oversize valve guides in plus 0.002 inch, 0.005 inch, 0.010 inch and 0.015 inch sizes are available as replacements. The valve guide to cylinder head interference should be within the range 0.0015 - 0.0025 inch. If there is ovality, necessitating the use of oversize valve guides, the bores in the head must be reamed oversize to suit. When new valve guides are fitted, it will be necessary to recut the valve seats.
11 When renewing the valve guides, do not omit the retaining circlips used only on the 850 cc.

20 Cylinder head - examination and renovation

1 Remove all traces of carbon from the combustion chambers and the inlet and exhaust ports, using a soft scraper which will not damage the surface of the valve seats. Finish by polishing the combustion chambers and ports with metal polish so that carbon does not adhere so readily. Never use emery cloth since the particles of abrasive will become embedded in the soft metal.
2 Check to make sure the valve guides are free from oil or other foreign matter that may cause the valves to stick.
3 If the valve seats are pocketed, as the result of excessive valve grinding in the past, the valve seats should be re-inserted. This is a specialist task which requires expert attention and quite beyond the means of the average owner. Pocketed valves cause a marked fall-off in performance and reduced engine efficiency as a direct result of the disturbed gas flow.
4 Make sure the cylinder head fins are not clogged with oil or road dirt, otherwise the engine may overheat. If necessary, use a wire brush but take care not to damage the light alloy fins which, in places, are thin in section.

21 Rocker arms and rocker spindles - examination and renovation

1 Examine carefully the outer surfaces of each rocker arm, to ensure there are no surface cracks or other signs of premature failure. The rocker arms should have a smooth surface, to resist any tendency towards fatigue failure.
2 The rocker arms should be a good sliding fit on the rocker spindles without excessive play. Noisy valve gear will result from worn rocker arms and spindles and performance may drop off as a result of reduced valve lift. If play is evident, the rocker arms should be renewed and new spindles fitted.
3 Check the rocker arm adjuster and the end of the rocker which engages with the pushrod. Both these points of contact have hardened ends and it is important that the surface is not scuffed,

chipped or broken, otherwise rapid wear will occur. The rocker adjuster and locknut can be renewed as separate items.
4 The rocker spindles must have a smooth, polished surface and an unobstructed oil way. Wear is most likely to occur if the flow of oil to the rocker gear is impeded in any way. For example, if the external rocker feed pipe unions are tightened carelessly, it is possible to 'neck' the thin pipe near the unions as the result of twisting action and seriously reduce the rate of oil flow.

22 Camshaft and pushrods - examination and renovation

1 The camshaft is unlikely to show signs of wear unless a high mileage has been covered or there has been a breakdown in the lubrication system. Wear will be most obvious on the flanks of the cams and at the peak, where flattening-off may occur. Scuffing, or in an extreme case, discoloration, is usually indicative of lubrication breakdown.
2 If there is any doubt about the condition of the camshaft, it is advisable to renew it whilst the engine is completely dismantled. Comparison with a new camshaft is often the best means of checking visually the extent of wear.
3 Check the pushrods for straightness by rolling them on a flat surface. Replace any that are bent, since it is impractical to straighten them with accuracy. Check that the hardened end pieces are not loose, or the internal bearing surfaces worn, chipped or broken.
4 Unless the machine is to be used for racing, no advantage is to be gained by fitting the Double S (Combat) or Triple S (Racer) camshafts. The standard 'Commando' camshaft will give a very high standard of performance with less mechanical noise and a lower rate of wear of the valve gear.

23 Engine reassembly - general

1 Before reassembly, the various engine components should be thoroughly clean and laid out close to the working area.
2 Make sure all traces of old gaskets and gasket cement have been removed and that the mating surfaces are clean and undamaged. One of the best ways to remove old gasket cement is to apply a rag soaked in methylated spirits. This acts as a solvent and will ensure the cement is removed without resort to scraping, with the consequent risk of damage.
3 Gather together all the necessary tools and have available an oil can filled with clean engine oil. Make sure all the new gaskets and oil seals are to hand, also any replacement parts required. There is nothing more infuriating than having to stop in the middle of a reassembly sequence because a vital gasket or replacement part has been overlooked.
4 Make sure the reassembly area is clean and well lit and that there is adequate working space. Refer to the torque and clearance settings, wherever they are given. Many of the smaller bolts are easily sheared if they are overtightened. Always use the correct size spanner and screwdriver, never an adjustable or grips as a substitute. If some of the nuts and bolts that have to be replaced were damaged during the dismantling operation, renew them. This will make any subsequent reassembly and dismantling much easier.
5 Above all else, use good quality tools and work at a steady pace, taking care that no part of a reassembly sequence is omitted. Short cuts invariably give rise to problems, some of which may not be apparent until a much later stage.

24 Engine reassembly - rebuilding the crankshaft assembly

1 Before the crankshaft is reassembled, check that all parts are clean and that the oilways are free, preferably by blowing through with compressed air. Arrange the parts in the correct order for reassembly, taking note of the alignment marks made when the crankshaft was dismantled.
2 Fit the left hand (drive side) crankshaft to the centre flywheel so that the ends of the two lower studs will locate with the right hand (timing side) crankshaft when both are mated up with the flywheel. A dowel in each face of the flywheel aids

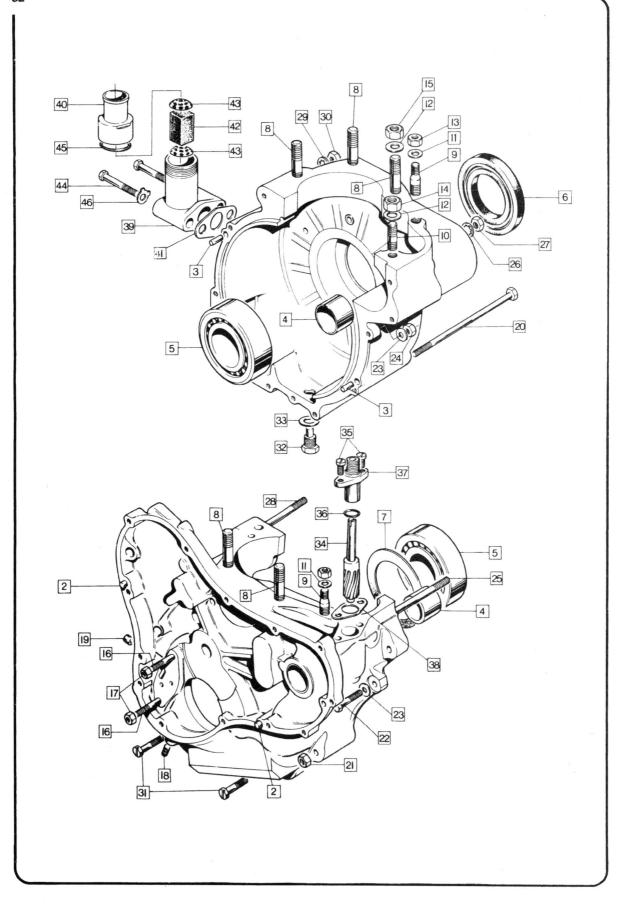

FIG. 1.6. CRANKCASE ASSEMBLY

1 Crankcase assembled (not supplied separately)
2 Crankcase dowel 2 off
3 Crankcase dowel 2 off
4 Camshaft bush 2 off
5 Main bearing 2 off
6 Main bearing oil seal
7 Main bearing shim
8 Cylinder stud 6 off
9 Cylinder stud stepped
10 Cylinder stud front
11 Cylinder stud washer 2 off
12 Cylinder stud washer 6 off
13 Cylinder stud nut 2 off
14 Cylinder stud nut
15 Cylinder stud nut 6 off
16 Oil pump stud 2 off
17 Oil pump stud nut 2 off
18 Oil stop grum screw
19 Dowel for junction block
20 Crankcase bolt, long
21 Crankcase bolt nut
22 Crankcase bolt short
23 Crankcase bolt washer 2 off

24 Crankcase bolt nut
25 Crankcase stud, top front
26 Crankcase stud washer
27 Crankcase stud nut
28 Crankcase stud top, rear
29 Crankcase stud washer
30 Crankcase stud nut
31 Crankcase screw 2 off
32 Magnetic sump plug
33 Sump plug washer
34 Tachometer drive gear
35 Screw 2 off
36 'O' ring seal
37 Tachometer drive housing
38 Gasket for tachometer drive housing
39 Breather body
40 Breather cap
41 Breather joint gasket
42 Separator block
43 Retainer disc 2 off
44 Bolt 2 off
45 'O' ring seal
46 Tab washer 2 off

22.1 Comparison of old and new camshafts, showing how wear occurs

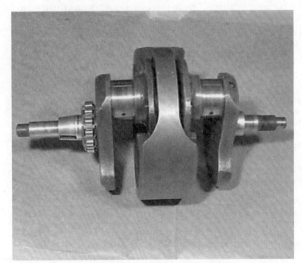

24.2 Locating dowels aid crankshaft reassembly

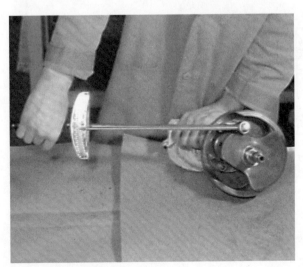

24.3 Tighten to 25 ft lb setting

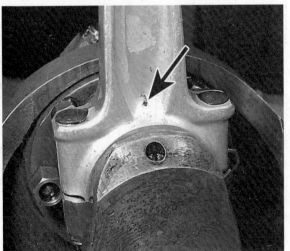

25.1 Connecting rod oilway must face OUTWARDS

location.

3   Insert the four remaining bolts from the left hand (drive) side and fit all six nuts. Note that the nuts fitted to the two lower studs should have a locking tab on each side of the flywheel, which must be bent over when the nuts are tightened fully. The four bolts may need to be tapped into position with a drift because they are a good fit in the crank flanges. Tighten all six nuts in a diagonal sequence, to a torque wrench setting of 25 lb ft. Check to ensure the oilways blanking plug is fitted in the right hand (timing side) crank cheek. This is important if a new right hand (timing side) crank cheek has been fitted. Centre-punch the nuts as a safeguard against slackening.

4   Pump oil through the crankshaft assembly with a pressure oil can to ensure all the oilways are clear and that the oil flow is not impeded in any way. A considerable amount of oil will be needed before it exudes from the oilways, due to the need to fill the area within the flywheel centre and both crank flanges.

25 Engine reassembly - refitting the connecting rods

1   Arrange the connecting rods, their end caps and their bearing shells in the correct order for reassembly, noting that these parts were marked, when dismantled, to ensure they are replaced in their original locations. The connecting rods MUST be fitted with the oilways from the big end bearing facing OUTWARDS.

2   Fit the shell bearing with the central oilways into the big end eye of each connecting rod so that the locating tab aligns with the depression in the connecting rod itself. Then fit the plain shell into each end cap, again locating the shell tab in similar fashion. Fit the connecting rods after oiling both bearing surfaces and the end caps, checking that the locating marks align and that the big end oilways in the connecting rods face OUTWARDS. Fit new connecting rod nuts and tighten them by hand. Check that the big end bearings are still free, then tighten the nuts with a torque wrench to a setting of 25 lb ft. Check again that the big end bearings are free from binding.

3   It is necessary to renew only the connecting rod nuts because it is the thread within the nuts which shows a tendency to stretch after a previous tightening and not the thread of the bolts.

26 Engine reassembly - reassembling the crankcases

1   Support the right hand crankcase open-end uppermost on the workbench and lower the crankshaft assembly into position.

Check that the right hand connecting rod is centrally disposed so that it will enter the crankcase mouth. A few taps with a rawhide mallet may be necessary, to ensure the right hand main bearing locates correctly.

2   When the crankshaft is fully home in the right hand crankcase, raise the assembly into the horizontal position and fit the oil sealing disc over the right hand end of the crankshaft, lipped side outwards. This should abut against the outer surface of the right hand main bearing. Fit the triangular-shaped washer, fit the Woodruff key into the crankshaft, then replace the crankshaft pinion so that the timing mark and chamfered teeth face outwards. It may be necessary to drive this pinion into position, using a tubular drift.

3   Lay the right hand crankcase on the workbench so that the crankshaft is again in a vertical position and fit the camshaft into the camshaft bush. Note that there is a thrust washer interposed between the camshaft and the bush on engines numbered up to 200000. Later engines employ two thrust washers in a similar position, the tabs of which must be correctly located. The end of the camshaft will engage with the tachometer drive pinion as the camshaft is inserted into the bush.

4   Lightly smear the mating surfaces of both crankcases with gasket cement. Insert the rotary breather disc and spring into the camshaft bush of the left-hand crankcase, retaining them with a dab of grease (early models only). Set the driven tangs of the disc so that they correspond with the slots in the camshaft end. Check that the left-hand connecting rod is correctly disposed in relation to the crankcase mouth, then lower the left-hand crankcase into position. Before the crankcases will join properly, it will be necessary to ensure that the breather disc has engaged in the camshaft end. When the two are correctly engaged, it will be possible to feel that they are positively connected by passing a piece of thin wire through the valve assembly and turning the camshaft. Having ensured that the valve, if fitted, is correctly aligned, the crankcases may be closed, using a hide mallet, if necessary.

5   Fit the front and rear screws which secure both crankcases together, also the short studs, nuts and washers. Check that both the crankshaft and the camshaft revolve freely without excessive end float, then fit the remainder of the crankcase retaining studs and bolts. Tighten the various nuts, bolts and screws evenly, in rotation, and again check that the crankshaft and camshaft revolve freely.

6   Fit a new oil seal in front of the left hand main bearing (drive side). It is a light drive-in fit.

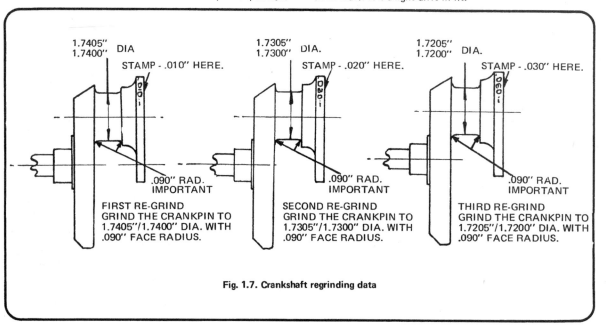

Fig. 1.7. Crankshaft regrinding data

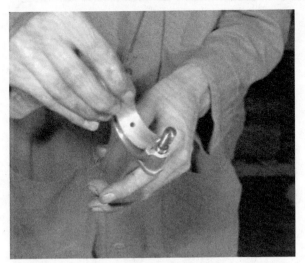
25.2 Make sure oilway in shell locates correctly

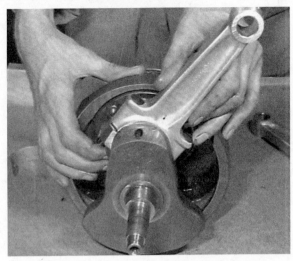
25.2a Always check bearings are free after final tightening

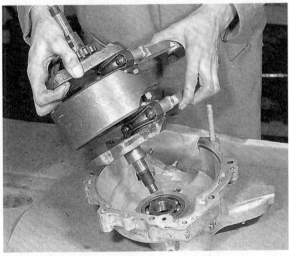
26.1 Fit crankshaft into right hand crankcase first

26.2 Fit the oil sealing washer first, then ...

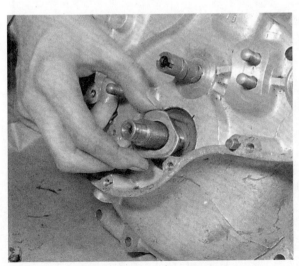
26.2a ... the triangular shaped washer, followed by ...

26.2b ... the crankshaft pinion, chamfered side outwards

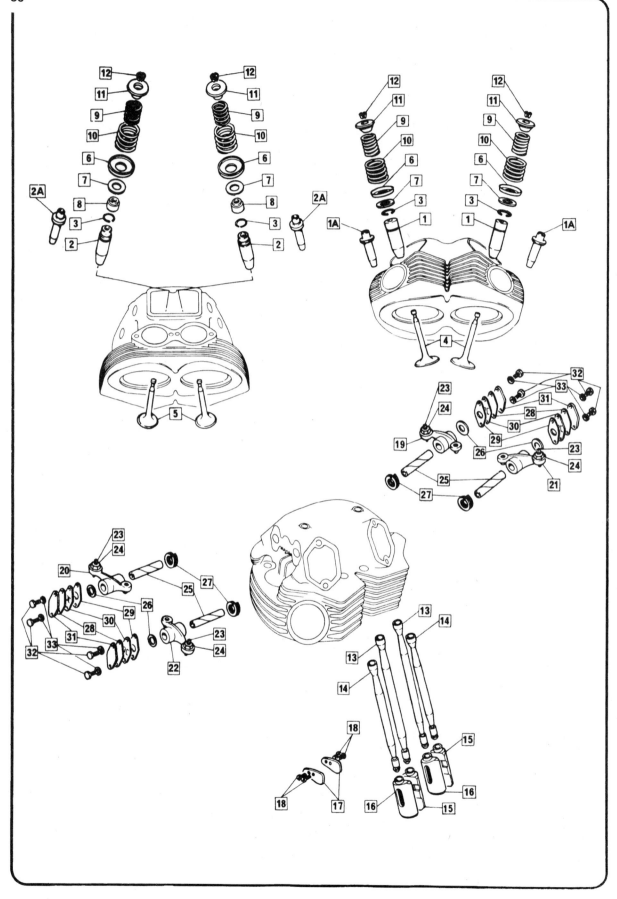

FIG. 1.8. VALVE GEAR

1   Valve guide, exhaust - 850 cc models 2 off
1a Valve guide, exhaust - 750 cc models 2 off
2   Valve guide, inlet - 850 cc models 2 off
2a Valve guide, inlet - 750 cc models 2 off
3   Circlip for valve guide (850 cc models only) 4 off
4   Exhaust valve 2 off
5   Inlet valve 2 off
6   Valve spring seat 4 off
7   Valve spring heat insulator* 4 off
8   Valve stem seal, inlet valves only 2 off
9   Valve spring, inner 4 off
10 Valve spring, outer 4 off
11 Valve collar 4 off
12 Split collet (pair) 4 off
13 Inlet pushrod 2 off
14 Exhaust pushrod 2 off
15 Tappet - left hand   )
16 Tappet - right hand) supplied in pairs only

17 Tappet locking plate 2 off
18 Tappet locking plate screw 4 off
19 Inlet rocker arm (left hand)
20 Inlet rocker arm (right hand)
21 Exhaust rocker arm (left hand)
22 Exhaust rocker arm (right hand)
23 Rocker adjuster 4 off
24 Rocker adjuster locknut 4 off
25 Rocker spindle 4 off
26 Rocker thrust washer 4 off
27 Rocker spring washer 4 off
28 Rocker plate gasket 4 off
29 Rocker spindle joint gasket 4 off
30 Rocker spindle lock plate 4 off
31 Rocker spindle retaining plate 4 off
32 Rocker retaining bolt 8 off
33 Copper washer 8 off
* Not fitted to inlet valves on RHI cylinder head

27 Engine reassembly - completing reassembly of the timing side

1   Turn the crankshaft assembly so that the timing mark on the camshaft pinion is uppermost (12 o'clock position). Loosely position the camshaft tensioner over its retaining studs, noting that the thinner of the two clamping plates slides over the studs first, long end downwards, followed by the tensioner arm, then the thicker clamping plate, long end uppermost. Fit, but do not tighten, the nuts and washers.

2   Assemble the intermediate gear pinion with integral sprocket, the camshaft sprocket and the camshaft chain. Both sprockets have a timing mark and must be positioned so that they are ten chain rollers apart. The gear pinion behind the chain sprocket has a timing mark, which must register exactly with the mark on the crankshaft pinion, with which it engages. A paint mark on the intermediate gear pinion in the vicinity of the timing mark aids location. When the timing marks register, slide the two camshaft chain sprockets onto their respective shafts, after inserting the Woodruff key into the camshaft to ensure correct register with the camshaft sprocket. It may be necessary to tap this latter sprocket into position.

3   Before continuing with the reassembly, again check the accuracy of the timing, to ensure all the timing marks are in correct register. It is easy to rectify any error at this stage.

4   Adjust the chain tensioner to give a maximum of 1/8 inch play at the tightest point of the chain run. Check the tension with the crankshaft in several different positions before tightening the chain tensioner nuts fully, to obviate the possibility of tight spots giving a false reading.

5   Fit the oil pump worm on the end of the crankshaft. The nut has a LEFT HAND thread. It can be tightened fully by locking the engine with a metal rod passed through the small eyes of both connecting rods.

6   Whilst the engine is still locked, fit the camshaft nut. This has a normal right hand thread.

7   Fit the oil pump using a new jointing gasket only, NO cement. Check to ensure the gasket aligns correctly with the oilways; if it is accidentally reversed, they may be masked off. The pump is secured by two nuts without washers, which must be tightened to a torque setting of 15 lb ft on models up to 1972; on later models be careful not to overtighten the nuts (maximum of 10 - 12 lb ft), or the studs will pull out of the crankcase.

8   If the oil pump has been dismantled, it must be primed with oil before fitting. This is accomplished by rotating it by hand whilst oil is fed into the gears with an oil gun.

9   Fit a new conical rubber seal over the oil pump outlet and dispense with any shims that may have been fitted previously. If the seal is over-compressed by the timing cover, permanent distortion will occur, with risk of leakage.

10 Check the tightness of the nut securing the oil pump driving gear.

11 Before refitting the timing cover, check that the mating surfaces are clean and undamaged and that both oil seals within the inner face of the cover are in good condition. If there is any doubt about the condition of these seals, they should be renewed as a precaution. The contact breaker seal is a push fit; the crankshaft oil seal is also a push fit but is retained by a circlip. See Section 16 of this Chapter.

12 Before fitting the timing cover, check that the blanking plug fitted to th ebody of the pressure release valve is located correctly.

13 A Norton Villiers service tool 061359 is available for fitting into the end of the camshaft, to permit the contact breaker oil seal to pass over the camshaft without damage as the timing cover is replaced. If the tool is not available, damage can be prevented by greasing the centre of the seal and the end of the camshaft and by taking special care to ensure correct alignment as the cover is fitted.

14 Lightly smear both mating surfaces with gasket cement, and use a new jointing gasket. Refit the timing cover, noting that the retaining screws are positioned according to their length, as shown in the accompanying illustration. Tighten them fully.

28 Engine reassembly - refitting the pistons and piston rings

1   Assemble the piston rings on each piston. The oil control rings are built up from three component parts, an expander and top and bottom rails.

2   Warm the pistons to aid insertion of the gudgeon pin and ensure one circlip is replaced in each piston boss, sharp edge outwards. The pistons are marked on the crown to ensure replacement in the correct position; the inlet valve cutaway is the one farthest away from the edge of the crown in each case.

3   Oil the small end eyes of the connecting rods and the gudgeon pins and gudgeon pin bosses. It is advisable to pad the mouth of the crankcase at this stage with clean rag, to prevent a misplaced circlip from falling in.

4   Support the piston and connecting rod, and press each gudgeon pin into position. Fit the second circlip in each case, sharp edge outwards. Make quite sure that BOTH circlips are correctly located with the groove inside each piston boss. A misplaced circlip will cause serious engine damage.

29 Replacing the crankcase assembly in the frame

1   It is convenient to replace the crankcase assembly in the frame at this stage, whilst the engine unit is still comparatively light in weight. Lift the crankcase assembly into position from the right hand side of the machine and locate the uppermost of the rear engine plate bolts first. This will steady the crankcase assembly whilst the remaining two bolts are inserted. Note that there is an electrical earthing connection taken from a tag washer fitted behind the nut of the lower rear engine plate bolt. THIS MUST NOT BE OVERLOOKED, otherwise the electrical system will be isolated by the Isolastic rubber suspension mountings.

2   Do not fit the front engine plates or front engine mounting at this stage. The greater angle of inclination will aid replacement of the cylinder head at a later stage.

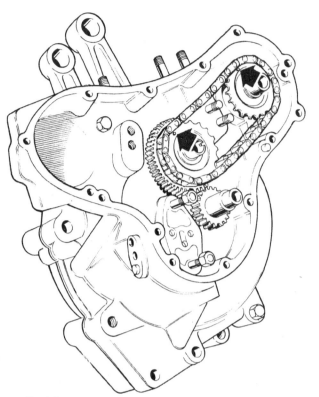

Fig. 1.9. Alignment of timing marks for correct valve timing.
Marks must be ten chain rollers apart

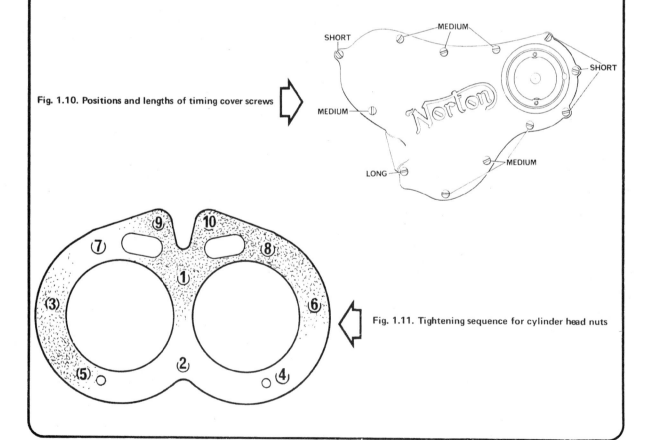

Fig. 1.10. Positions and lengths of timing cover screws

Fig. 1.11. Tightening sequence for cylinder head nuts

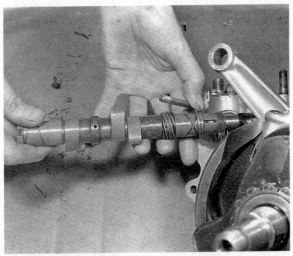
26.3 Fit camshaft into right hand crankcase, not omitting thrust washers

26.4 Early models have engine breather disc driven by camshaft

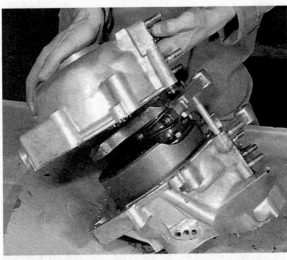
26.4a Make sure end of camshaft locates with breather disc when assembling crankcases together

26.6 Fit a new oil seal in front of left hand main bearing

27.1 Assemble timing chain tensioner first, but do not tighten

27.2 Camshaft sprocket is keyed on to camshaft to ensure correct location

27.2a Drive sprocket on to camshaft, after ...

27.2b ... fitting timing sprockets and chain in unison. Note timing marks

27.5 Oil pump worm is retained by nut with LEFT HAND thread

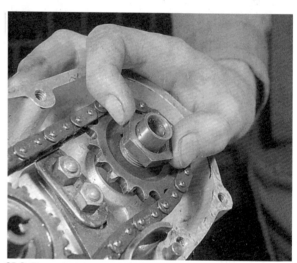

27.6 Fit and tighten camshaft sprocket nut whilst engine is locked

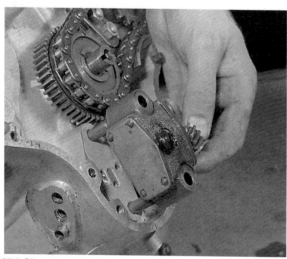

27.7 Oil pump slides into position down mounting studs

27.9 Fit new conical rubber to pump outlet

27.14 Use gasket cement at timing cover joint, with new gasket

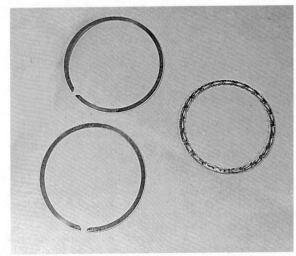

28.1 Oil control rings are built up from an expander and two 'rails'

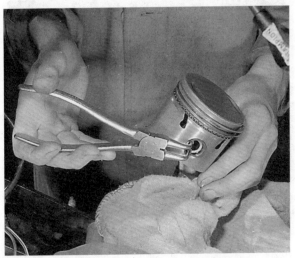

28.3 Pad crankcase mouth to trap any displaced parts

28.4 If pistons are warmed, gudgeon pins will press into position

29.1 Do not **omit** electrical connection, or system will be isolated

30.2 Piston ring clamps make fitting barrels much easier

30 Engine reassembly - refitting the cylinder barrel

1 Position the piston rings so that their end gaps do not coincide and so that the end gap of the oil control ring is clear of the cutaway in the base of each cylinder bore. The rail gaps should be at least one inch to either side of the expander gap, when the ring is built up. If not correctly positioned, the rail ends may become trapped with the cylinder base cutaway and break up.

2 Smear both pistons and the cylinder bores with clean engine oil, then fit a pair of piston ring compressors to each piston. Fit a new cylinder base gasket (no gasket cement). Place a pair of wooden or metal rods below each piston, across the crankcase mouth, to steady the pistons as the cylinder barrel is lowered into position.

3 When the piston rings have engaged with the bores, remove the clamps and the rods below the pistons. At the same time, withdraw the rag used for padding the crankcase mouth.

4 Before the cylinder barrel is lowered fully, engage the cylinder base nuts, since there is insufficient clearance for them to be fitted when the barrel is seated on the crankcase mouth. Only the rear centre nut lacks a washer. The four additional through bolts on the 850 cc should have new sealing washers fitted before tightening down.

5 Tighten the cylinder base nuts in sequence. The 3/8 inch nuts and bolts should be tightened to a torque setting of 25 lb ft and the 5/16 inch nuts to 20 lb ft.

31 Engine reassembly - reassembling the cylinder head

1 Replace the valves and valve springs, together with their associated seatings and collars etc. Do not omit the valve stem seals, fitted to the inlet valves only.

2 If the rocker spindles have to be replaced, heat the cylinder head to 150° - 200°C before they are driven back into position. Each rocker arm should have a double spring washer fitted against the innermost boss and a plain washer against the other boss, to take up end float. The outer end of the rocker spindle must be aligned so that the two slots are in a horizontal position before it is driven flush with the jointing face. The flat on the spindle must face the rocker cover in each case.

3 Oil the rocker spindles and rocker arms after reassembly and check that all four rocker arms move freely, without binding. Refit the end covers which are secured by two bolts. There is no necessity to renew the gaskets, unless there has been a previous leakage, or unless the locking plate and retaining plate have been separated.

32 Engine reassembly - refitting the cylinder head

1 Place a new cylinder head gasket on top of the cylinder barrel and turn the engine so that the pistons are at top dead centre (TDC) to reduce valve lift to a minimum. There are three types of cylinder head gasket. The type fitted will depend on the desired compression ratio.

2 Insert the pushrods in the pushrod tunnels of the cylinder head with the longer pushrods in the innermost position on each side. The cupped end of the pushrods should face upwards. Feed the pushrods into the cylinder head as far as possible.

3 Invert the cylinder head and whilst holding the pushrods with one hand, position the cylinder head over the cylinder barrel. Allow the pushrods to drop into the tunnels of the cylinder barrel, where they will locate with the cam followers.

4 Lower the cylinder head and engage the ends of the pushrods which are slightly higher with the ball ends of the rockers with which they engage. Some skill is necessary during this operation, using a screwdriver or a piece of stout wire to guide the pushrods into position through the exhaust rocker cover orifice. When they have engaged correctly, tackle the second pair in similar fashion.

5 Tighten down the cylinder head using the sequence detailed

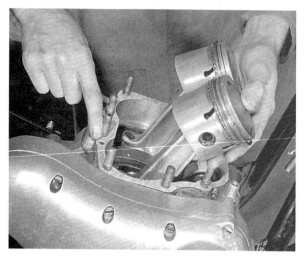

30.2a Check hole in base gasket aligns with oilway

30.2b Metal rods support pistons while barrel is lowered

30.4 Engage base nuts and washers before barrel is lowered completely

32.4 Do not tighten down until the push rods have engaged correctly

32.5 Tighten down the front centre cylinder head bolt first, as shown

32.5a Do not overlook the nuts fitted from the front underside

32.6 Order of assembly of the thrust washers and shims etc

32.6a It will be necessary to ease the mounting into position

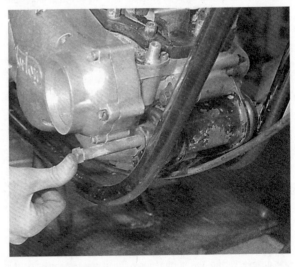

32.6b Fit centre bolt, making sure head clears crankcase castings

in the accompanying illustration. The short cylinder head bolt and washer should be tightened down first to overcome the spring pressure of the valves on lift and cause the head to seat squarely on the head gasket. Check all four pushrods have engaged correctly before commencing the tightening sequence. The torque settings are:

3/8 inch nuts and bolts	30 lb ft
5/16 inch bolts	20 lb ft

6 Refit the front engine plates and the front engine mounting by reversing the dismantling procedure. It will be necessary to raise the engine during this operation. Replace the crankcase breather pipe.

7 A Britool 3/8 in. W. square drive socket spanner used in conjunction with the Britool A97 extension and a free tommy bar, is recommended for tightening the recessed cylinder head bolts. A Stanwille or Gordon ¼ in. W. open box spanner is recommended for the other bolts that have restricted access.

33 Engine reassembly - reassembling the primary transmission

1 Replace the Woodruff key used to locate the engine sprocket on the left hand end of the crankshaft. Lightly smear the smooth circular face of the left hand crankcase with gasket cement and the matching face of the back of the inner chaincase casting. Fit a new gasket to the crankcase face and fit the inner chaincase, taking care that the gearbox mainshaft does not damage the oil seal within the centre of the rear of the casting. It is advisable to grease both the seal and the mainshaft to obviate risk of damage.

2 Check that the chaincase is seating squarely and that the centre stud which passes through the chaincase has the correct number of spacing washers. If the spacers are absent or of incorrect thickness, there will be a tendency for the chaincase to bow when the centre nut is tightened and give rise to oil leaks.

3 Replace the three bolts which retain the chaincase to the crankcase. Each should have a tab washer fitted, to lock the bolt in position after it has been tightened fully. Failure to observe this precaution may cause a bolt to work loose and jam the primary transmission. Do not omit to bend the locking tabs.

4 Slide the clutch location spacer over the gearbox mainshaft, recessed portion inwards. Add the spacing washers, necessary to ensure correct alignment of the sprockets and chain.

5 The triplex chain has no spring link and it is necessary to fit the engine sprocket, chain and clutch as a unit. Before assembling and fitting these components, check the centre bearing of the clutch. This is a ball journal bearing, retained by a circlip. If any play is evident or if the bearing runs roughly as the clutch body is rotated, it should be replaced. It is a drive fit in the clutch body.

6 Fit the clutch, engine sprocket and chain over the respective shafts. The clutch has a splined centre and the engine sprocket has a keyway which engages with the key already inserted in the crankshaft. If necessary, use a tubular drift to ensure both sprockets are correctly located to the full depth of engagement. Fit the clutch centre securing nut and spring washer or nut and tab washer, depending on the year of manufacture of the model concerned. Apply the rear brake by slipping a tube over the operating arm, or by temporarily reconnecting the brake pedal, so that the clutch centre can be locked whilst the centre nut is tightened fully. A torque setting of 70 lb ft is necessary. Do not omit to bend the tab washer (if fitted).

7 Reassemble the clutch plates. The plain steel plate with the two small dowels on the inner face must be fitted first; the dowels engage with matching holes in the main body of the clutch. Then fit the friction and steel plates in alternate order, ending with the extra thick steel plate which has a serrated outer rim. Later models may have some variation of the clutch make-up including the use of sintered bronze friction plates in place of the earlier friction material used and a specially-hardened clutch centre. The method of assembly is, however, broadly identical.

8 The clutch diaphragm should be reassembled with the compressor tool and tensioned so that it is completely flat. Push the diaphragm, complete with compressor tool, as far into the clutch as possible and enter one end of the retaining circlip into the groove within the clutch body. Feed the remainder of the

circlip into the groove and check that it has located correctly BEFORE RELEASING THE COMPRESSOR. This precaution cannot be overstressed. If an attempt is made to fit the diaphragm without the correct type of compressor, or if the retaining circlip is not located positively, THERE IS RISK OF PERSONAL INJURY if the compressed diaphragm works free.

9 Replace the clutch pushrod within the hollow mainshaft, after coating it with grease. Insert it through the centre of the clutch, and replace the pushrod adjuster screw and locknut. Adjust the pushrod by slackening off the handlebar adjuster completely and screwing in the adjuster until the clutch commences to lift. Slacken back the adjuster one complete turn and lock it in this position with the locknut. It may be necessary to detach the inspection cover from the gearbox during this operation because if the clutch operating arm within the gearbox outer shell has dropped out of location, clutch action will be rendered inoperative. It can be raised back into position if the clutch adjuster is temporarily slackened off.

10 Replace the shims on the end of the crankshaft, in front of the engine sprocket, and the spacer, recess facing outwards. This will enable the Woodruff key which locates the alternator rotor to be inserted in the end of the crankshaft. Clean the rotor to ensure the magnetism has not attracted any metallic particles and fit the rotor over the keyway, with the name and timing mark facing outwards. Replace the rotor nut and shaped washer, then engage top gear and apply the rear brake so that the engine is locked whilst the nut is tightened to a torque setting of 70 - 80 lb ft.

11 Fit the three spacers on the rotor mounting studs followed by the stator coil assembly. The lead from the stator assembly should be positioned in the five o'clock position, projecting from the surface which faces outwards. This will ensure the lead is kept well clear of the primary chain. Pass the lead through the orifice in the inner chaincase, using the rubber grommet to make an effective oil seal. The snap connector terminals can be reconnected with the snap connectors above the gearbox.

12 Replace the plain washers and stator retaining nuts. Tighten the nuts to a torque setting of 15 lb ft. There must be a minimum air gap between the stator coil assembly and the rotor of from 0.008 - 0.010 inch and a check should be made with a feeler gauge. If the gap is reduced at any point, misalignment of the stator mounting studs should be suspected and corrected.

13 Fit a new sealing band within the recessed portion of the inner chaincase facing, then attach the outer chaincase, which is secured by the single centre sleeve nut and washer. Tighten the nut fully. It is advisable to refill the chaincase at this stage as a check against oil leakage and to prevent this operation being overlooked. It holds 200 cc of engine oil.

14 Refit the left hand footrest and brake pedal assembly, which is secured by two nuts and washers, and a bolt. Note that the support carrying the stop lamp switch cables is fitted over the bottom stud and acts in lieu of a washer. Reconnect the rear brake cable (if disconnected previously) and adjust the cable so that only minimal movement of the brake pedal is necessary to apply the brake. Reconnect the stop lamp switch and adjust the height of the switch if necessary.

34 Engine reassembly - completion

1 Refit the twin carburettors complete with their spacers and heat insulators. Do not overtighten the socket screws which retain the spacers and carburettors to the cylinder head, or there is risk of the flanges bowing and causing air leaks. Replace the slide and needle assembly in each carburettor, taking care that the needle enters the needle jet and that the air slide locates with its slot in the jet block. Replace the top of each mixing chamber, which is retained by two crosshead screws. Check that the twist grip and air control lever operate smoothly, without any of the carburettor components jamming.

2 Replace the air cleaner element and the end and side covers. The complete assembly is held together by two long bolts, one on each side of the air cleaner unit. Reconnect the short rubber hoses with the carburettor intakes.

3 Replace both spark plugs, after checking to ensure they are gapped correctly and not fouled.

32.6c Re-fit crankcase breather pipe

33.1 Fit new gasket to inner chaincase joint with crankcase

33.1a Ease chaincase oil seal over gearbox mainshaft

33.4 Check gearbox mainshaft spacers are located correctly

35.5 Clutch centre bearing is a drive fit

33.5a Clutch centre bearing is retained by a circlip

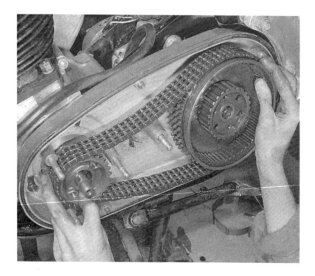

33.6 Fit engine sprocket, chain and clutch sprocket in unison

33.7 Clutch plate with dowel pins is inserted first

33.7a Friction plates are one-piece

33.7b End assembly with thick dished plate

33.8 Improvised compressor for clutch diaphragm

33.8a Make sure circlip is fully located before releasing compressor

33.9 Grease and insert push rod

33.9a Fit adjuster and adjust as described in text

33.10 Recessed end of spacer must face outwards to permit fitting of key

33.10a Replace rotor with timing marks outwards

33.11 Alternator lead must face OUTWARDS to clear chain

33.13 'O' ring around chaincase forms effective oil seal

33.14 Attach guide for stop lamp wires to footrest stud

a 0.006 inch feeler gauge will be a good sliding fit (0.008 inch, Combat engine).

3   If the clearance is not correct, slacken the locknut on the end of the rocker arm and turn the square end of the adjuster until the correct setting is achieved. Hold the square end of the adjuster steady and tighten the locknut, then recheck the clearance. If correct proceed to the left hand inlet valve, which is checked in similar fashion whilst the right hand inlet valve is open fully.

4   Use the same technique to check and if necessary adjust the two exhaust valves. The right hand valve must be open fully when the left hand valve clearance is checked, and vice versa. Note that in the case of the exhaust valves, the correct clearance is 0.008 inch (0.010 inch, Combat engine). Always recheck after any adjustment is made.

5   When the correct settings have been obtained for all four valves, fit new gaskets and replace the two exhaust and one inlet rocker covers. No gasket cement should be necessary if the jointing faces are in good condition and the covers tightened fully.

6   Replace both spark plugs and reconnect the plug caps.

4   Refit the cylinder head steady. The baseplate is retained by three socket screws which should not be tightened fully until the top mounting plates and bolts have been located and tightened. The head steady assembly must be tight.

5   Refit the twin ignition coils by reversing the dismantling procedure and connect the spark plug caps.

6   Fit the auto-advance unit and contact breaker cam to the end of the camshaft, within the contact breaker housing. Do not push on to the taper at this stage. Fit the contact breaker plate and points assembly and replace but do not tighten the two screws or hexagonal pillars which lock the baseplate in position. Feed the twin lead from the contact breaker points through the tunnel at the side of the contact breaker housing and rejoin the snap connectors with the lead from the coils assembly. Set the ignition timing by following the procedure given in Chapter 5.9.

7   Refit the exhaust system by reversing the dismantling procedure. A new sealing washer should be fitted within each exhaust port. Ensure both finned exhaust pipe clips are tightened fully - if they work loose the internal threads of the cylinder head will be shattered away. Later models use split clamps within the finned rings and have special tab washers to lock the rings in position. It is best to use Norton Villiers service tool 063968 for tightening the finned rings without damage, or a 'C' spanner of the appropriate diameter. The 850 cc has a balance pipe connecting both exhaust pipes, immediately below the point at which the pipes emerge from the cylinder head. Do not omit to tighten the clamps.

8   Reconnect the rocker feed pipe to the rear of the timing cover and to the cylinder head by means of the banjo unions provided. Use new copper sealing washers at the union joints. Refit the oil pipe junction block, using a new gasket at the joint (no gasket cement), and tighten the retaining bolt firmly. Replace the pressure release valve in the rear of the timing cover, if it has been removed for cleaning, and tighten to a torque setting of 25 lb ft. Check that the crankcase drain plug and/or filter plug has been replaced and tightened, then refill the oil tank with 5 Imperial pints of new engine oil.

34.1 Overtightening socket screws will bow flanges

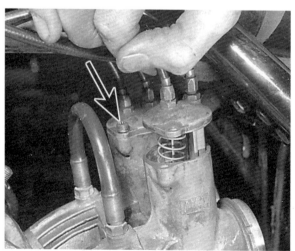

34.1a Top of mixing chamber is secured by two cross head screws

35 Checking and resetting the valve clearances

1   In order to rotate the engine more easily, remove both spark plugs. The spark plugs were fitted at an earlier stage to preclude the possibility of washers being dropped into the engine whilst reassembling.

2   Rotate the engine until the left hand inlet valve is open fully, then check the clearance of the right hand inlet valve. This check should be made with the engine cold. If the clearance is correct,

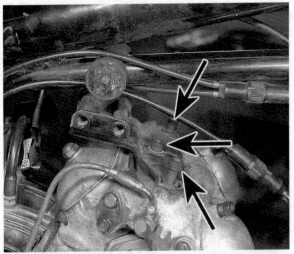

34.4 Base of head steady is retained by three socket screws

34.4a Tighten side plates fully before tightening base plates

34.5 Ignition coil assembly faces rearwards

34.6 Fit auto-advance unit on camshaft taper

34.6a Replace the contact breaker base plate but do not tighten fully

34.7 Place a new sealing ring in each exhaust port

34.7a Make sure silencer mountings are tight

34.7b Special 'C' spanner is advisable for tightening exhaust rings

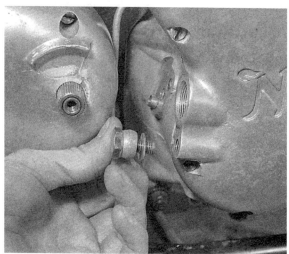

34.8 Rocker feed pipe connects at rear of timing cover

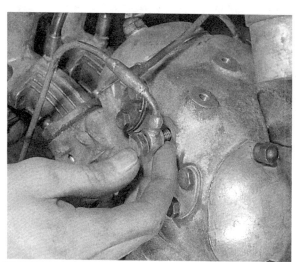

34.8a Ensure unions do not twist whilst tightening

34.8b Use new gasket at oil pipe block joint

34.8c Replace pressure release valve and tighten to 25 ft lb

35.2 Check all valve clearances

not necessarily mean improved performance; in a great many instances unwarranted modifications or the fitting of an unsuitable design of silencer will have an adverse effect on both performance and petrol consumption.

37 Increasing engine performance

1 Norton Villiers Limited manufacture a number of conversion kits for improving the performance of the 750 Racer and 750 Formula Racer, which can in some cases be applied to the standard road models. Details of these kits, which include ancillary parts such as fairings, tanks and handlebars etc are available through any Norton Villiers agent. Advice should always be sought before any modification work is undertaken since some parts are not recommended for use on the standard road models.

36 Starting and running the rebuilt engine

1 Replace the petrol tank and reconnect the fuel lines. Do not replace the seat at this stage, since it is necessary to check whether the oil is returning to the oil tank when the engine is started. Refill the oil tank.
2 Replace the fuse in the fuse holder, fitted to the battery negative lead. Check that the electrical system is functioning correctly by means of the lights and the ammeter reading.
3 Switch on the ignition and start the engine. Run it at fast tick-over speed until oil commences to return to the oil tank. There may be a time lag before the flow of oil issues from the return pipe because pressure has to build up in the rebuilt engine before circulation is complete. Do not permit the engine to run at low speed for more than a couple of minutes without evidence of the oil returning, before stopping it and checking the lubrication system. To verify whether the oil pump is working, slacken the pressure relief valve a little and see whether oil emerges from around the threads, when the engine is restarted.
4 If the engine refuses to start, despite evidence of a good spark and petrol in the carburettor, try changing over the plug leads. It is easy to transpose them, especially if the engine was timed using a different procedure.
5 The exhaust will smoke excessively during the initial start, due to the presence of excess oil used during the reassembly of the various components. The smoke should clear gradually, as the engine settles down.
6 The return to the oil tank will eventually contain air bubbles because the scavenge pump will have cleared the excess oil content of the crankcase. The scavenge pump has a greater capacity than the feed pump, hence the presence of air when there is little oil to pick up.
7 Refit the seat and check the engine for leakages at gaskets and pipe unions etc. It is unlikely any will be evident if the engine has been reassembled correctly, with new gaskets and clean jointing faces. Before taking the machine on the road check that both brakes work effectively and that all controls operate freely.
8 If the engine has been rebored, or if a number of new parts have been fitted, a certain amount of running-in will be required. Particular care should be taken during the first 100 miles or so, when the engine is most likely to tighten up, if it is overstressed. Commence by making maximum use of the gearbox, so that only a light loading is applied to the engine. Speeds can be worked up gradually until full performance is obtained with increasing mileage.
9 Do not tamper with the silencer or fit another design unless it is designed specifically for a Norton twin. A noisier exhaust does

36.1 Don't forget to refill oil tank, also ...

36.1a ... the primary chaincase

38 Fault diagnosis

Symptom	Cause	Remedy
Engine will not start	Earthed cut-out button	Temporarily disconnect lead and check for spark at plugs
	Reversed plug leads	Transpose leads. (Likely to occur during rebuilds.)
	Contact breaker points closed	Check and re-adjust points.
	Flooded carburettor	Check whether float needle is sticking and clean.
Engine runs unevenly and misfires	Incorrect ignition timing	Check setting and adjust if necessary.
	Faulty or incorrect grades of spark plug	Clean or replace plugs.
	Fuel starvation	Check fuel lines and carburettor.
Lack of power	Incorrect ignition timing (retarded)	Check and reset timing. Check action of automatic advance unit.
Engine pinks	Incorrect ignition timing (over-advanced)	Check and reset timing.
	Compression ratio too high (tuned engines only)	Replace pistons if running on top grade fuel.
Excessive mechanical noise	Worn cylinder block (piston slap)	Rebore and fit O/S pistons.
	Worn small end bearings (rattle)	Replace bearings and gudgeon pins.
	Worn big end bearings (knock)	Replace shell bearings and regrind crankshaft.
	Worn main bearings (rattle)	Fit new bearings.
Engine overheats and fades	Lubrication failure	Check oil pump and oil pump drive.

Chapter 2 Gearbox

Contents

Specifications

Ball journal bearings

Mainshaft (clutch end)	1¼ x 2½ x 5/8 in
Mainshaft (kickstarter end)	5/8 x 9/16 x 7/16 in
Layshaft (clutch end)	17 mm x 40 mm x 12 mm

Gearbox pinions - number of teeth

Layshaft	4th	14 teeth	Mainshaft	4th	23 teeth
Layshaft	3rd	20 teeth	Mainshaft	3rd	21 teeth
Layshaft	2nd	24 teeth	Mainshaft	2nd	18 teeth
Layshaft	1st	28 teeth	Mainshaft	1st	14 teeth

Gearbox bushes - dimensions

4th gear bush	0.9060 - 0.9053 in (23.012 - 22.995 mm) outside diameter
	0.8145 - 0.8140 in (20.688 - 20.675 mm) inside diameter
	0.8133 - 0.8120 in (20.657 - 20.625 mm) inside diameter fitted
Layshaft bush	1.126 - 1.124 in (28.60 - 28.549 mm) inside diameter
Mainshaft 2nd gear bush	0.8125 - 0.8115 in (20.637 - 20.608 mm) inside diameter fitted
Layshaft 3rd gear bush	0.8125 - 0.81 in (20.637 - 20.57 mm) inside diameter fitted
Layshaft 1st gear bush	0.6885 - 0.6875 in (17.488 - 17.462 mm) inside diameter fitted
Footchange spindle bush	0.6290 - 0.6285 in (15.976 - 15.964 mm) inside diameter fitted
Camplate/quadrant bush	0.5005 - 0.4995 in (12.713 - 12.687 mm) inside diameter fitted
Kickstarter spindle bush	0.675 - 0.673 in (17.145 - 17.094 mm) inside diameter fitted

Camplate plunger spring free length ... ...   1.500 in (38.1 mm)

Primary transmission

Engine sprocket	26 teeth
Clutch chainwheel	57 teeth
Gearbox sprocket	19 teeth (750 cc standard size)
	21 teeth (850 cc standard size)
Primary chain	0.375 x 0.250 in, triplex

Gear ratios

		750 cc models			850 cc models
Gearbox sprocket size	19	20	21	21
4th gear	4.84 : 1	4.60 : 1	4.38 : 1	4.38 : 1
3rd gear	5.90 : 1	5.60 : 1	5.30 : 1	5.30 : 1
2nd gear	8.25 : 1	7.80 : 1	7.45 : 1	7.45 : 1
1st gear	12.40 : 1	11.80 : 1	11.20 : 1	11.20 : 1

Gearbox capacity (oil) ... ...   0.75 Imp pints (0.9 US pints/420 cc) SAE 90 EP Oil

1 General description

The gearbox fitted to all Norton 750 cc and 850 cc
Commando models is of the four-speed positive footchange
type, virtually identical to the AMC gearbox first introduced by
this latter Company in 1957. The only exception relates to the
750 Production racer and the 750 Racer which are fitted with a
Quaite five-speed close ratio gearbox. These models are not
normally used on the road.

The gearbox is a self-contained unit and it is not necessary to
dismantle or remove the engine in order to gain access. If it is
desired to remove the gearbox from the frame, only the primary
transmission need be dismantled. The rear engine plates are
designed to release the gearbox after the rear wheel and rear
engine bolts have been removed.

Most repair work can be accomplished with the gearbox in
position, although it will be necessary to drain the oil and remove
the inner and outer covers if the gear clusters, selector arms or
camplate require attention.

2 Dismantling the gearbox - removing the inner and outer covers

1 To remove the gear change lever, first unscrew the centre
screw which retains the gear indicator in position. Then slacken
the pinch bolt through the gear change lever and draw the lever
off its splined shaft.
2 Slacken and remove a similar pinch bolt through the base of
the kickstarter and draw the kickstarter off its splined shaft. In
this instance it is necessary to withdraw the pinch bolt
completely since it locates with a groove between two sets of
splines.
3 If the battery is still connected, remove the fuse so that the
electrical system is isolated.
4 Remove the two nuts and single bolt which secure the right
hand footrest to the light alloy mounting plate. Note that the
red-coloured earth lead from the electrical system is attached to
the rear of the mounting plate, beneath the nut which is on the
end of the single bolt.
5 Place an oil drain tray below the gearbox and remove the
drain plug. Allow the oil content to drain off whilst proceeding
with the next operations.
6 Unscrew and remove the five cheese-head screws which secure
the gearbox end cover and the two screws which retain the outer
cover inspection cap. The inspection cap can be lifted away,
together with its sealing gasket.
7 Detach the end of the clutch cable from the fork of clutch
operating lever, within the gearbox outer cover. Access is through
the inspection cap orifice.
8 Position the oil drain tray so that it will catch any residual
oil which may be released when the outer cover is worked loose
and withdrawn. The ratchet plate and spindle will most
probably remain attached to the outer cover; if not, withdraw
them from the inner cover housing.
9 The clutch withdrawal lever assembly is free to revolve in its
housing when the locking ring is removed. Mark the body and
the locking ring with punch marks so that they can be aligned
correctly on reassembly.
10 Remove the screw and nut retaining the clutch withdrawal
lever and detach the lever, roller and roller sleeve which are
displaced. This operation is necessary only if wear or breakage
is suspected.
11 Unscrew the locking ring around the clutch withdrawal lever
assembly and withdraw the body complete with the clutch
operating ball.
12 Select top gear by levering the end of the gear change
quadrant towards the top of the aperture in the gearbox inner
cover, whilst rocking the rear wheel to facilitate the engagement
of the gear pinion dogs. When top gear is engaged, apply the rear
brake fully and unscrew the mainshaft nut. This has a normal
right hand thread.
13 Lever the gear change quadrant in the opposite direction

until neutral is located, then unscrew and remove the seven nuts
which retain the gearbox inner cover. Two of these nuts are on
the outside of the casing, at the base of the gearbox. Remove the
inner cover by lightly tapping the end of the mainshaft to aid
separation. The cover can now be lifted away. There is no
necessity to disturb the kickstarter spindle, pawl and return
spring assembly unless attention to these parts is required. They
will remain attached to the inner cover.

3 Dismantling the gearbox - removing the gear clusters and camplate

1 Before further dismantling can take place, it is necessary to
remove the clutch from the end of the mainshaft and the final
drive sprocket from the sleeve gear. Refer to Chapter 1, Section
8 for the appropriate procedure. The engine can be locked
during the dismantling operation by selecting top gear and
applying the rear brake, preferably before the gearbox inner
cover is removed.
2 The final drive sprocket is secured by a large diameter nut
with a LEFT HAND thread and a locking plate and grub screw.
Remove the grub screw and locking plate, then prevent the
sprocket from turning whilst the nut is slackened by passing the
final drive chain around it and securing both ends in a vice. A
1½ inch AF set spanner is needed to fit the nut.
3 If not already removed from the mainshaft, take off any
clutch location shims and the clutch locating spacer. They are
retained by a circlip. Reverting to the other end of the gearbox,
withdraw the low gear pinion from the end of the mainshaft
and unscrew the selector fork spindle. The selector forks can now
be disengaged from the camplate and withdrawn, then the clutch
pushrod from within the mainshaft, followed by the mainshaft
itself complete with gear cluster.
4 Withdraw the layshaft complete with gear cluster, then
remove the mainshaft sleeve gear. It is a tight fit in the gearbox
main bearing and should be displaced by driving it inwards into
the gearbox shell, through the main bearing, using a rawhide
mallet to obviate the risk of damage. The sleeve gear oil seal and
spacer will be displaced during this operation.
5 Unscrew and remove the acorn-shaped nut from the front
underside of the gearbox shell. This contains the spring-loaded
camplate plunger which will be exposed when the nut is
removed. Remove the two bolts and washers which secure the
camplate and quadrant to the gearbox shell and lift these
components out. Do not lose the knuckle pin roller in the end of
the quadrant assembly.

4 Dismantling the gearbox - removing the gearbox bearings

1 The gearbox has four bearings, two ball journal bearings in
the left hand side of the gearbox shell to support the mainshaft
sleeve gear and layshaft respectively. The gearbox inner cover
contains a further ball journal bearing for the right hand end of
the mainshaft; the right hand end of the layshaft is supported
in a bush fitted within the kickstarter shaft.
2 The ball journal bearings and the bush are an interference fit
within their respective housings. It is necessary to heat the
housing in each case so that the bearing can be drifted out of
position, or if the housing is blind, by bringing the housing
down sharply on a flat wooden surface so that the bearing is
displaced by the shock. It is essential that heat is applied before
any attempt is made to remove a bearing.

5 Removing the gearbox as a complete unit

1 Occasions may arise where it is desirable to remove the
gearbox as a complete unit, without disturbing the engine. As
mentioned earlier it will be necessary to dismantle the primary
transmission as a prelude to the removal of the gearbox, as
detailed in Chapter 1.8.

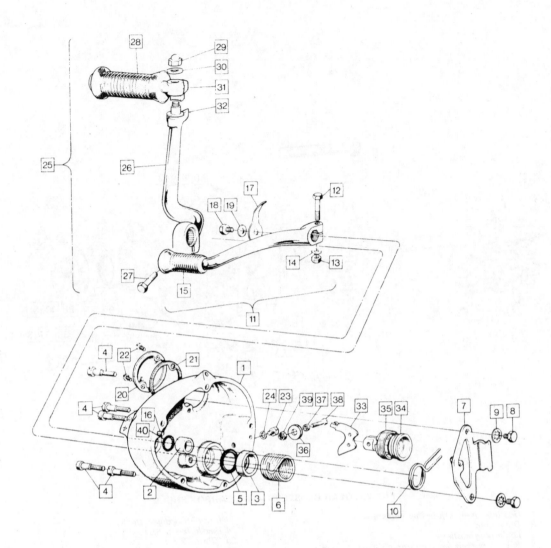

FIG. 2.1. GEARBOX OUTER COVER ASSEMBLY

1   Gearbox outer cover
2   Gearchange lever bush
3   Kickstarter lever bush
4   Outer cover screw 5 off
5   Kickstarter shaft 'O' ring
6   Kickstarter return spring
7   Gearchange stop plate
8   Gearchange stop plate bolt 2 off
9   Gearchange stop plate washer 2 off
10 Gearchange return spring
11 Gearchange lever
12 Gearchange lever bolt
13 Gearchange lever nut
14 Gearchange lever washer
15 Gearchange lever rubber
16 Dowel 2 off
17 Gearchange indicator
18 Gearchange indicator bolt
19 Gearchange indicator bolt washer
20 Inspection cover

21 Inspection cover gasket
22 Inspection cover screw 2 off
23 Oil level bolt
24 Oil level bolt washer
25 Kickstarter lever
26 Kickstarter crank only
27 Kickstarter pinch bolt
28 Kickstarter rubber
29 Domed nut
30 Plain washer
31 Pin
32 Spring washer
33 Clutch operating lever
34 Clutch operating lever body
35 Clutch operating lever body lock ring
36 Clutch operating roller
37 Clutch operating roller sleeve
38 Clutch operating roller screw
39 Clutch operating roller screw nut
40 'O' ring for pawl carrier

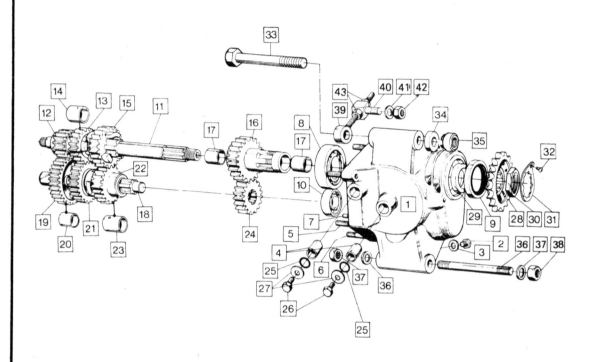

FIG. 2.2. GEAR CLUSTERS AND GEARBOX SHELL

1 Gearbox shell with bushes and studs
2 Drain plug
3 Drain plug washer
4 Bush, quadrant and cam spindle 2 off
5 Stud for inner case 2 off
6 Stud for inner case 5 off
7 Dowel 2 off
8 Sleeve gear bearing
9 Sleeve gear bearing oil seal
10 Layshaft bearing
11 Mainshaft
12 Mainshaft 1st gear
13 Mainshaft 2nd gear
14 Mainshaft 2nd gear bush
15 Mainshaft 3rd gear
16 Sleeve gear complete with bushes
17 Sleeve gear bush 2 off
18 Layshaft
19 Layshaft 1st gear with bush
20 Layshaft 1st gear bush
21 Layshaft 2nd gear
22 Layshaft 3rd gear

23 Layshaft 3rd gear bush
24 Layshaft 4th gear
25 'O' ring 2 off
26 Spindle bolt 2 off
27 Spindle bolt washer 2 off
28 Gearbox sprocket (19 - 24 teeth sizes available)
29 Gearbox sprocket spacer
30 Gearbox sprocket nut
31 Sprocket nut lockwasher
32 Gearbox sprocket nut lockscrew
33 Gearbox top bolt
34 Gearbox top bolt spacer
35 Gear top bolt nut
36 Gearbox pivot stud
37 Gearbox pivot stud washer 2 off
38 Gearbox pivot stud nut 2 off
39 Gearbox adjuster
40 Gearbox adjuster crosshead
41 Gearbox adjuster crosshead washer
42 Gearbox adjuster crosshead nut
43 Gearbox adjuster nut

2.6 Remove inspection cap for access to clutch cable

2.8 Ratchet plate and spindle will remain attached to outer cover

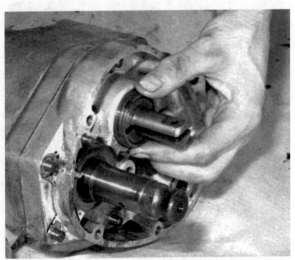
2.9 Mark withdrawal lever assembly to aid reassembly in same location

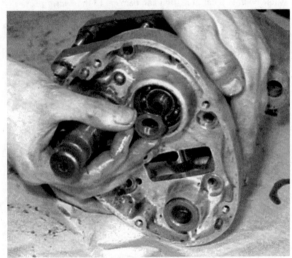
2.12 Gearbox mainshaft nut has right hand thread

2.13 Remove inner cover with gearbox in neutral

3.2 Unscrew the grub screw ...

2   When the primary chaincase has been removed completely, place a stout wooden box below the lower frame members, to support the machine as the centre stand is detached. Remove the rear wheel and the three bolts (or studs) which retain the rear of the crankcase to the rear engine plates. Detach the top and bottom gearbox mounting bolts and remove the draw bolt which serves as the gearbox position adjuster from the right hand side of the rear engine plates. Working from the right hand side of the machine, turn the gearbox anticlockwise in the engine plates. Force the rear engine plates rearwards until the cutaway of the bottom right hand side is clear of the crankcase. It is now possible to turn the gearbox further, so that it can be withdrawn horizontally through the aperture in the right hand engine plate.

6 Examination and renovation - general

1   Each of the various gearbox components should be examined carefully for signs of wear or damage after they have been cleaned thoroughly with a petrol/paraffin mix. A cleansing compound, such as Gunk or Jizer, is particularly useful if the gearbox castings are covered with a film of oil and grease. Make sure all the internal parts have been removed if the gearbox has been dismantled prior to treatment, otherwise the subsequent water wash will cause rusting and damage to the bearings.
2   All gaskets and oil seals should be renewed, regardless of their condition, if the rebuilt gearbox is to remain oiltight. A rag soaked in methylated spirits provides one of the best means of removing old gasket cement, without having to resort to scraping and risk of damaging the mating surfaces.
3   Check for any stripped studs or bolt holes which must be reclaimed before reassembly. Internal threads in castings can often be repaired cheaply by the use of what is known as a 'Helicoil' thread insert, without need to tap oversize. Many motor cycle repairers can offer a 'Helicoil' service.

7 Gear pinions, selector arms and bearings - examination and renovation

1   Examine each of the gear pinions to ensure there are no chipped, rounded or broken teeth and that the dogs on the ends of the pinions are not rounded. Worn dogs are a frequent cause of jumping out of gear; renewal of the pinions concerned is the only effective remedy. Check that the inner splines are in good condition and the pinions are not slack on the shafts. Bushed pinions require special attention in this respect, since wear will cause them to rock.
2   Check both the layshaft and the mainshaft for worn splines, damaged threads and other points at which wear may occur, such as the extremeties which pass through the bearings. If signs of binding or local overheating are evident, check both shafts for straightness.
3   Examine the selector forks to ensure they are not twisted or badly worn. Wear at the fork end will immediately be obvious; check the arm in conjunction with the gear pinion groove with which it normally engages. Do not overlook the pin which engages with the camplate track; this is subject to wear.
4   The three ball journal bearings should be washed out with petrol and examined for damaged ball tracks. Reject any bearing which has more than just perceptible side play, or is noisy or rough when rotated. Check the bush within the kickstarter spindle which should be renewed if the layshaft is a slack fit.

8 Kickstarter ratchet and gear selector mechanism - examination and renovation

1   The kickstarter ratchet is cut within the outer face of the layshaft bottom gear pinion and is engaged by means of a spring-loaded pawl attached to the kickstarter shaft. After a lengthy period of use, the edges of the ratchet teeth will wear in conjunction with the edge of the pawl and the kickstarter will

3.2a ... remove lock washer. Nut has LEFT HAND thread

3.3 Clutch spacer and shims are retained by a circlip

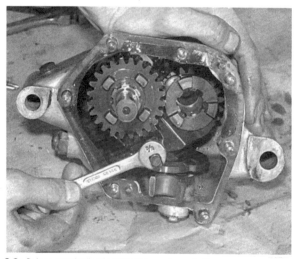

3.3a Selector spindle screws into gearbox shell

show a tendency to slip under heavy load. If the ratchet teeth show signs of wear on their leading edges, renewal of the bottom gear pinion will be necessary, also the pawl. It is a wise precaution to renew the pawl spring on the same occasion, especially since it is a low cost item.

2 Check the kickstarter return spring, which is coiled around the kickstarter shaft, has not weakened or stretched. It is a wise precaution to renew this spring too, whilst the gearbox is dismantled.

3 The gear change mechanism rarely gives trouble, apart from the occasional breakage of the gear change lever return and/or ratchet springs. Both springs should be examined and replaced if there is any doubt about their condition.

4 The gearbox camplate should be examined, especially if the pins of the selector forks have worn. The tracks in the camplate are subject to wear after an extended period of service, wear that will be most obvious where the tracks change direction. Wear in the selector mechanism will render gear changes less precise and will cause a general sloppiness of the gear change lever.

5 Do not overlook the camplate plunger and spring. If the plunger jams in its housing or the tension spring weakens, the gearbox will tend to jump out of gear, since the gear which is engaged will no longer be retained positively. Make sure that the acorn nut which forms the housing for the plunger and spring is tightened fully, otherwise some of the spring pressure will be lost.

9 Gearbox shell and end covers - examination and renovation

1 Examine the gearbox shell and the inner and outer covers for cracks or damaged mating surfaces. Small cracks will require expert attention, since they can probably be repaired by welding. Larger or more extensive cracks will necessitate the purchase of a replacement. It is not practicable to effect a satisfactory repair to a badly cracked casting in view of the distortion that may occur.

2 Damaged mating surfaces will cause oil leaks and if the indentations or marks are deep, renewal of the casting will be necessary. Small imperfections can often be sealed off if a liberal coating of gasket cement is used in the area affected, in combination with a new gasket. Beware of oil leakage from the gearbox since the oil may reach the rear tyre and cause the rear wheel to lose adhesion with the road surface. If in doubt, always play safe and renew the defective part.

10 Gearbox reassembly - refitting the gearbox bearings, gear clusters and selectors

1 Heat the gearbox shell in order to expand the bearing housings and press both the mainshaft sleeve gear bearing and the layshaft bearing into their correct locations. Make sure both bearings enter their housings squarely and are driven fully home.

2 Working from the outside of the gearbox, fit a new oil seal into the bearing housing, lipped side towards the sleeve gear bearing.

3 Locate the gear change quadrant within the gearbox shell bearing and replace the retaining bolt and washer, which should be tightened fully. Do not omit the O ring seal which must be replaced, if damaged.

4 Raise the knuckle of the quadrant until the uppermost end of the curved portion is in line with the top right hand stud of the gearbox shell (see accompanying illustration). Whilst the quadrant is retained in this position, fit the camplate so that the teeth of the quadrant engage with the teeth at the rear of the camplate. The smooth edge of the camplate should face outwards and the last indentation in the other portion of the camplate should be positioned immediately above the orifice of the camplate plunger. If the quadrant and camplate are lined up in this position, the gear change mechanism is 'timed' correctly and the gears will select in the correct sequence.

5 Replace the bolt and washer which retain the camplate in position, not omitting the O ring seal. Replace the camplate

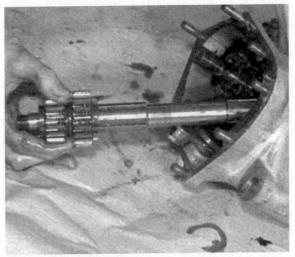
3.3b Withdraw mainshaft first

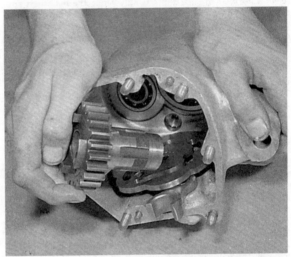
3.4 Sleeve gear must be drifted through main bearing from outside

3.5 Camplate assembly is secured to gearbox shell by two bolts

plunger and tighten the acorn nut fully. Check that the alignment of the camplate in relation to the quadrant is still correct.

6 Fit the sleeve gear, complete with bushes, through the sleeve gear bearing and oil seal, taking care not to damage the latter. Coat the inside of the oil seal and the projecting shaft of the sleeve gear to obviate risk of damage, whilst the sleeve gear is driven through the bearing from within the gearbox shell.

7 Fit the spacer for the final drive sprocket and then the final drive sprocket itself. The sprocket has a splined centre fitting which engages with the end of the sleeve gear shaft and is retained by a large diameter nut with a LEFT HAND thread. Tighten to a torque setting of 80 lb ft and fit the locking plate and screw. The sprocket can be held steady during the tightening operation by wrapping the final drive chain around it and holding both ends of the chain in a vice.

8 Insert the mainshaft through the sleeve gear pinion and fit the layshaft third gear pinion and bush to the layshaft, followed by the layshaft fourth gear pinion. Fit the fourth gear pinion with its flat face against the third gear pinion and its shouldered face towards the bearing. The layshaft, complete with pinions, can now be pushed into the layshaft main bearing, within the gearbox shell.

9 Assemble together the mainshaft third gear pinion and the selector fork, then slide the assembly along the mainshaft and engage the pin of the selector fork with the innermost track of the camplate. Replace the mainshaft second gear, complete with centre bush, on the mainshaft. The dogs on the end of this pinion should face the inside of the gearbox.

10 Assemble together the layshaft second gear pinion and selector fork, slide the assembly along the layshaft and engage the pin of the selector fork with the outer track of the camplate. The selector rod can now be inserted through both selector forks and threaded into the gearbox shell. The rod has a flat on the outer end so that it can be tightened fully with a spanner.

11 Fit the layshaft first gear pinion and the mainshaft bottom gear pinion, the latter with its long extension facing outwards.

11 Gearbox reassembly - refitting the inner cover

1 Before refitting the inner cover of the gearbox, make sure the roller is replaced in the end of the gear change quadrant arm. It is not possible to fit the roller after the inner cover has been located since projections in the casting prevent its insertion.

2 If the mainshaft ball journal bearing has been displaced for renewal, the inner cover casting must be heated before the new replacement is fitted. Check that the kickstarter spindle, if removed, is replaced so that the pawl is behind the stop on the inner cover, as shown in the accompanying illustration.

3 Replace the kickstarter return spring on the kickstarter shaft. The spring is positioned so that the projecting portion faces outwards; the spring should be slid down the kickstarter shaft until the forward facing end enters the locating hole in the shaft. Check that the pawl is located correctly, as detailed in the preceding paragraph, then tension the kickstarter spring by turning the curved end of the spring clockwise until it locates with the stop pin to the right of the kickstarter shaft housing.

4 Lightly smear the jointing faces of the gearbox shell and the inner cover with gasket cement and fit a new jointing gasket. Ensure the dowel pins are in position in the gearbox shell.

5 Check that all the gearbox components are pushed home fully, especially the selector rod, which must engage to the full depth of thread. Then fit the inner cover, checking that it locates with the dowels and that the quadrant roller is still in position. It may be necessary to guide the unsupported end of the selector rod into position and to twist the inner cover to and fro a small amount so that the dowels will locate correctly. When the cover is fully home, replace the seven nuts which retain the end cover, but before they are tightened, check that both the layshaft and the mainshaft revolve quite freely. If they bind or if the inner cover will not seat correctly, one of the gearbox components has not located correctly or has been assembled in incorrect order.

If the end cover is tightened under these conditions it may crack or become permanently distorted.

6 When the gearbox shafts revolve freely, tighten the seven nuts which retain the inner cover, to a torque setting of 10 - 15 ft lb.

7 Replace the mainshaft nut on the right hand end of the gearbox mainshaft and lock the gearbox so that it can be tightened to a torque setting of 70 ft lb. If the gearbox has been dismantled whilst in the frame, reconnect the final drive chain, select top gear and apply the rear brake to lock the mainshaft during the tightening operation. If the gearbox is out of the frame, wrap the chain around the final drive sprocket and clamp both ends in a vice to form an equally effective lock.

8 If the clutch withdrawal lever assembly has been dismantled, which is rarely necessary, reassemble the lever, roller, bush and pivot screw and tighten the locknut. Grease and replace the clutch pushrod, then locate the clutch withdrawal lever and locking ring, not forgetting to insert the large diameter ball bearing first. This bears directly on the pushrod end and is actuated by the clutch withdrawal lever.

9 Tighten the locking ring, whilst the clutch withdrawal lever assembly is held in its correct location. The centre punch marks made when the assembly was dismantled will ensure correct re-location. Tighten the locking ring fully.

12 Gearbox reassembly - refitting the outer cover

1 If the gear change mechanism within the outer cover has been dismantled, a check should be made to ensure the spring retaining washer is fitted between the pawl carrier assembly and the outer cover itself. The pawl carrier assembly must be able to move freely and the ratchet spring must be located correctly. The outer cover can be fitted with the ratchet plate assembly attached, or inserted into the already reassembled inner cover; the method of reassembly is identical.

2 Lightly smear the mating surfaces of the inner and outer end covers with gasket cement and fit a new jointing gasket. Check that the locating dowels are fitted to the inner cover.

3 Locate the ratchet spring in the central position and guide the outer cover over the kickstarter shaft, so that it engages with the dowels. If the cover shows any reluctance to locate fully, it is probable that the pawl has rotated, because the ratchet spring

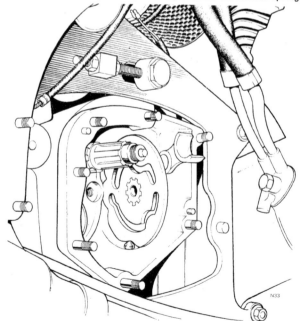

Fig. 2.3. When reassembling gearbox, knuckle end of quadrant must be in line with the top right hand stud of the gearbox shell. This will ensure correct 'timing' of gears

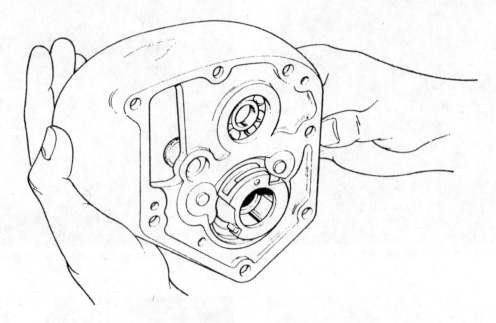

Fig. 2.4. Kickstarter pawl stop must locate with inside of inner cover, as shown

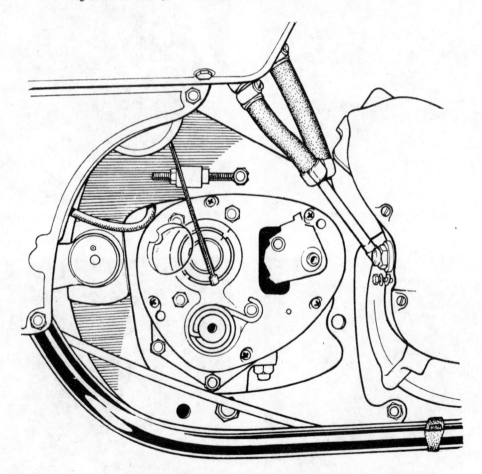

Fig. 2.5. Clutch body must be aligned to give straight pull on cable. Note location of kickstarter spring

5.2 Top gearbox bolt has special flat head

5.2a Turn gearbox sideways to free from specially-shaped engine plates

7.1 Comparison of new and damaged gear pinions

10.9 Fit mainshaft first, then layshaft and build up within shell

10.10 Layshaft second gear and selector fork are fitted together ...

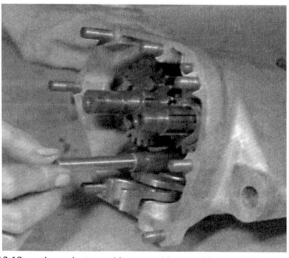

10.10a ... then selector rod is screwed into position and tightened fully

10.11 Fit layshaft bottom gear pinion, then ...

10.11a ... mainshaft bottom gear pinion, extension facing OUT-WARDS

11.1 Roller must be replaced now. It cannot be inserted later

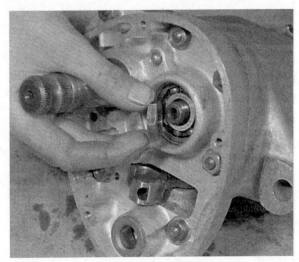

11.7 Tighten gearbox mainshaft nut to 70 ft lb

11.8 Do not omit large diameter ball bearing

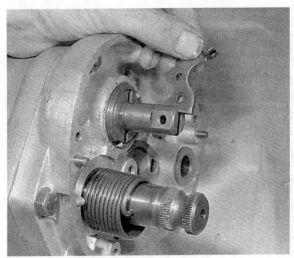

11.8a Fit clutch withdrawal lever, then ...

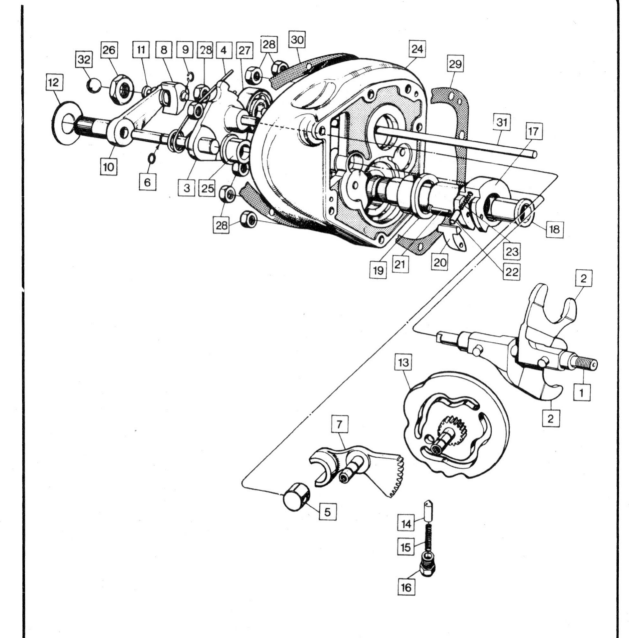

FIG. 2.6. GEARBOX INNER COVER ASSEMBLY

1 Selector fork spindle
2 Selector fork 2 off
3 Ratchet plate assembly
4 Ratchet spring
5 Knuckle pin roller
6 Ratchet spindle 'O' ring
7 Quadrant
8 Gearchange pawl
9 Gearchange pawl circlip
10 Pawl carrier assembly
11 Pawl pivot pin
12 Spring washer
13 Camplate
14 Cam plunger
15 Plunger spring
16 Plunger spring bolt

17 Kickstarter shaft with bush
18 Kickstarter shaft bush
19 Inner cover bush for kickstarter
20 Kickstarter pawl
21 Kickstarter pawl pin
22 Kickstarter pawl plunger
23 Kickstarter pawl spring
24 Gearbox inner cover
25 Gearchange inner bush
26 Mainshaft nut
27 Mainshaft bearing
28 Gearbox inner cover stud nut
29 Gasket, inner cover to gearbox shell
30 Gasket, inner cover to outer cover
31 Clutch pushrod
32 Clutch operating ball

was not located correctly and has permitted the pawl to move. It will be necessary to withdraw the outer cover and reset the pawl and spring before making another attempt at refitting.

4 When the inner cover is fully home, replace and tighten the five screws which retain the cover in position. Refit the kickstarter and gear change lever and check that both operate correctly. It is important that the pinch bolts of both levers are tightened fully, otherwise the levers will work loose and damage the splines. If the pinch bolt is tight and yet the lever is a slack fit on the splines, the joint which is pulled together by the pinch bolt can be opened up with a hacksaw in which two or three blades have been fitted to give a wider cut.

5 Check the gear selecting action by engaging each gear in turn whilst rotating the rear wheel to facilitate the engagement of the dogs. If the gears select in a positive and satisfactory manner, check that the gearbox drain plug has been refitted and is tightened fully, then refill the gearbox with oil through the clutch lever inspection cover. It holds 0.75 Imperial pints (420 cc) of SAE 90EP oil. Reconnect the clutch cable, making sure the nipple engages correctly with the clutch withdrawal lever, then replace the inspection cover, using a new gasket. The cover is secured by two screws.

6 Before taking the machine on the road for the initial test run, verify that the clutch adjustment is correct. It is most important that there is no loading on the clutch pushrod, otherwise overheating will cause the hardened ends of the rod to soften and adjustment to require frequent attention.

13 Primary chain - examination, lubrication and adjustment

1 Although the primary chain runs under near ideal conditions, in the sense that it is totally enclosed and is running in an oil bath, it will nonetheless require periodic attention to take up any slackness that has developed as the result of wear.

2 Adjustment is effected by drawing the gearbox backwards in the rear engine plates, so the distance between the centres of the engine sprocket and the clutch sprocket is increased. A threaded inspection cap in the primary chaincase provides a convenient means of checking that the chain tension is correct and of examining the chain if it is suspected that any damage has occurred.

3 Chain tension is correct if there is a movement of 3/8 inch in the middle of the top run of the chain, when it is at its tightest point. It is advisable to measure the amount of play when the engine is in several different positions, because a chain rarely wears in an even manner. If adjustment is necessary, slacken the top and bottom gearbox locating bolts and use the drawbolt adjuster on the right hand side, immediately above the gearbox, to draw the gearbox backwards the appropriate amount. Retighten the gearbox bolts and again check the chain tension.

4 After an extended period of service, the chain will wear to the extent where replacement is required. Note that it will be necessary to remove the clutch sprocket and the engine sprocket in unison with the chain, as described in Chapter 1.8, before the chain can be measured for wear. It must also be washed free of oil with a petrol/paraffin mix.

5 To check for wear, compress the chain endwise so that all play in the rollers is taken up. Make a mark at each end, then anchor one of the ends and pull the chain in the opposite direction so that all play is taken up in the opposite direction. If the chain extends by more than ¼ inch per foot, it is due for renewal. When renewing the chain, it is advisable to renew the engine and clutch sprockets on the same occasion, otherwise a more rapid rate of wear will occur as the result of old and new parts being run together.

6 Occasionally, the chain should be removed, washed and then immersed in a special chain lubricant such as Linklyfe or Chainguard, which has to be heated before the chain can be immersed and then lifted out to dry off. This type of lubricant gives better penetration of the chain rollers and is not thrown off so readily by the rotary motion of the chain.

7 If the machine is used at irregular intervals, or for a series of short journeys, the oil in the primary chaincase should be changed at more frequent intervals than recommended by the routine maintenance instructions. This will help offset the effects of condensation, a rusty brown deposit on the chain links which indicates the oil is becoming contaminated.

8 Use only a good quality chain. Although cheap chains are available they have a much shorter working life in many cases. Remember that a broken chain can cause extensive damage, apart from the possibility of locking the primary transmission without warning.

14 Changing the gearbox final drive sprocket

1 Although the manufacturer has selected gear ratios that give optimum performance and it is therefore difficult to improve overall performance as the result, occasions will arise when a change has to be made, such as when engine performance is increased by tuning. Norton Villiers can supply a range of gearbox final drive sprockets for the standard road models, ranging in size from 19 teeth to 24 teeth.

2 From the preceding Sections of this Chapter, it will be appreciated that a certain amount of dismantling is necessary in order to effect a sprocket change, namely the removal of the primary transmission and chaincase, as described in Chapter 1.8. In consequence, a change in sprocket size is not a matter to be taken lightly, especially since the whole process will need to be repeated if the end result is not satisfactory. It is therefore advisable to seek the advice of a Norton Villiers specialist before contemplating such a change; the fitting of a larger final drive sprocket by no means guarantees an increase in maximum speed.

3 The gearbox final drive sprocket should be examined at intervals to check for wear, chipped or broken teeth or other defects which will otherwise cause rapid chain wear and harsh transmission. A badly worn sprocket should be renewed at the earliest possible opportunity, preferably in conjunction with the chain and the rear brake drum which has an integral sprocket.

11.8b ... roller and bush, before tightening the pivot screw

12.5 Refill with SAE 90EP oil

15 Fault diagnosis

Symptom	Cause	Remedy
Kickstarter does not return when engine is turned over or started	Broken kickstarter return spring	Replace.
Kickstarter slips and will not turn engine over	Worn ratchet assembly	Replace.
Kickstarter jams	Worn ratchet assembly and/or pawl	Replace.
Difficulty in engaging gears	Gear selectors not indexed correctly	Check alignment of 'timing' marks on inner cover.
	Selector forks bent or badly worn	Replace.
Machine jumps out of gear	Camplate plunger sticking	Remove and free.
	Worn dogs on gear pinions	Replace defective pinions.
Gear change lever does not return to original position	Broken return spring	Replace spring in outer cover.

Chapter 3 Clutch

Contents

Specifications

Clutch

Type ... ...   Multi-plate
Number of friction plates   4
Thickness of friction plates   ... ...   0.148 in - 0.142 in
Centre bearing   ... ...   62 mm o/d, 35 mm i/d, 14 mm wide deep groove ball bearing type, one dot
Pushrod length ... ...   9.813 - 9.803 in (249.250 - 248.996 mm)
Pushrod diameter ... ...   0.237 - 0.232 in (5.984 - 5.857 mm)

Clutch adjuster

Diameter ... ...   ½ inch
Thread size ... ...   20 UNF 2A

1 General description

A multi-plate clutch with a hardened steel centre and a large diameter diaphragm spring is fitted to the Norton Commando. This design of clutch, although somewhat unusual on motor cycles, has the advantage of dispensing with the separate clutch springs used on conventional clutches and the need to adjust their tension so that the clutch plates lift evenly. More important, it is possible to exert strong spring pressure by the use of a diaphragm spring, whilst still retaining light hand operation.

2 Dismantling the clutch

1   Access to the clutch is gained by removing the outer cover of the primary chaincase which is retained by the centre sleeve nut. Since the chaincase does not have a drain plug, a container should be placed under the chaincase before the sleeve nut is slackened. When the inner and outer chaincases are parted, approximately 200 cc of oil will be released.

2   If only the clutch plates require attention, there is no necessity to dismantle the primary transmission. The clutch plates are readily accessible with the clutch chainwheel in situ; they can be withdrawn after the clutch diaphragm spring has been compressed and removed.

3   It is essential to use Norton Villiers service tool 060999 to compress the diaphragm spring, or alternatively to make up a compressor as illustrated in Fig 3.1. Any attempt to compress the diaphragm without the appropriate tool is extremely hazardous. IF THE TENSION OF THE COMPRESSED SPRING IS RELEASED WITHOUT PROTECTION, SERIOUS INJURY MAY RESULT.

4   Screw the centre bolt of the compressor tool into the clutch

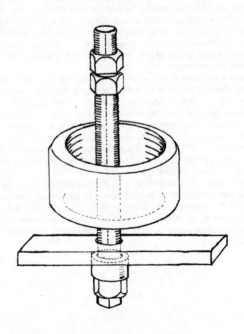

Fig. 3.1 Home made clutch compressor

centre, after the clutch adjuster and locknut have been removed. At least ¼ inch of thread must engage with the internal thread in the clutch centre. Tighten the withdrawal nut until the diaphragm spring is flat and will revolve within the clutch housing. The retaining circlip can now be prised out of the groove within the periphery of the clutch chainwheel; use a screwdriver to prise out the first end, then 'peel' the circlip from its location. When the circlip has been removed, the diaphragm spring and compressor can be lifted away as a pair; there is no necessity to separate them prior to reassembly.

5   To withdraw the clutch plates, make a wire hook and use it to pull the plates from the body of the clutch. They can be drawn out quite easily.

3 Clutch plates - examination and renovation

1   Check each clutch plate to ensure it is completely flat and free from any blemishes. Reject any that are distorted, or clutch troubles will persist.

2   Examine the teeth at the edge of the plain clutch plates and in the centre of the friction plates. It is important that they should be in good condition and free from any burrs or other damage. Burrs can be removed by dressing with a file; if any teeth are chipped or broken, the clutch plate should be replaced.

3   The thickness of each friction plate should be measured and compared with the Specifications limits given at the beginning of this Chpater. Clutch slip at high speeds is usually an indication that the thickness of the friction plates is at the lower limit. Always replace the friction plates as a complete set, never singly.

4   Clutch slip will also occur if the early type friction plates are soaked in oil. The sintered bronze plates eliminate this trouble but MUST be used in conjunction with the specially-hardened clutch centre.

4 Clutch inner and outer drums - examination and renovation

1   Examine the slots with which the clutch plate teeth engage. If they are indented badly, renewal of the inner and/or outer drum will be necessary. Minor indentations can be dressed with a file so that the slots have parallel sides again. Clutch drag and general uncertainty of clutch action can usually be attributed to such indentations which trap the plates and restrict their freedom of movement.

2   Indifferent clutch action can be attributed to a worn centre bearing. Wear is easily detected by gripping the clutch body and attempting to rock it sideways. If excess movement is evident, the centre bearing must be renewed. Under these circumstances it is necessary to remove the clutch, engine sprocket and primary chain as one complete unit. Refer to Chapter 1.8, paragraphs 10 and 11, for the recommended procedure. Note that if the clutch centre nut has slackened, this can give the impression that the bearing has failed.

3   The clutch chainwheel should be examined, checking for general wear, chipped or broken teeth. If damage of this nature is found, the complete clutch outer drum must be renewed since the chainwheel is an integral part. The same reference to the dismantling procedure as that in the preceding paragraph will apply.

5 Clutch diaphragm spring - examination

The clutch diaphragm spring rarely gives trouble and can be expected to have a very long life: The point at which wear is most likely to occur is at the tips of the 'fingers' which bear the compressive load. Excessive wear at this point will be self-evident.

6 Reassembling the clutch

1   Reassemble the clutch by reversing the dismantling procedure detailed in Section 2 of this Chapter. Make sure that the clutch plates are alternated correctly and that the pins in the back of the innermost plain plate locate correctly with the main body of the clutch. Do not forget to replace the clutch pushrod, which may have been withdrawn during any dismantling operation.

2   When replacing the diaphragm spring, ensure that the circlip has entered its groove correctly and is positively located before the compressor tool is slackened off. Failure to observe this precaution may result in personal injury if the pressure is released suddenly and the circlip has not seated correctly. Replace the clutch adjuster and locknut, after positioning a dab of grease on the end of the adjuster bolt where it makes contact with the pushrod. Adjust the clutch according to the procedure recommended in the following paragraph.

7 Clutch adjustment - general

1   In order to adjust the clutch correctly, the following procedure must be followed. Commence by slackening the handlebar lever adjuster off completely, so that there is no tension on the clutch cable. Then screw in the adjuster in the clutch centre until it makes contact with the end of the pushrod and there is just perceptible movement of the clutch diaphragm. Slacken the adjuster bolt back one complete turn and hold it in this position whilst the locknut is tightened. Then re-adjust at the handlebar lever until there is about 1/16 inch to 3/32 inch free play before the cable is under tension.

2   It is most important that the amount of free play in the clutch cable is maintained. If the clutch pushrod is permanently in contact with the clutch actuating mechanism, the continuous loading will cause the rod to heat up and the hardened ends soften. Rapid wear of the pushrod will follow, necessitating frequent clutch adjustment.

3   When making the initial adjustment, occasions occur when the clutch withdrawal lever is displaced, rendering the clutch inoperative. Under these circumstances it is necessary to remove the inspection cover from the right hand side of the gearbox, slacken back the adjuster in the clutch centre and lift the lever back into its correct location before adjusting the clutch again. When replacing the inspection cover, do not omit the sealing gasket.

4   There is no necessity to detach the outer portion of the chaincase in order to effect clutch adjustment. A threaded circular inspection cover in the centre of the clutch dome permits access to the adjuster, without need to drain even the oil content.

8 Fault diagnosis

Symptom	Cause	Remedy
Engine speed increases but not road speed	Clutch slip; incorrect adjustment or worn linings	Adjust or replace clutch plates.
Machine creeps forward when in gear; difficulty in finding neutral	Clutch drag; incorrect adjustment or damaged clutch plates	Re-adjust or fit new clutch plates.
Machine jerks on take-off or when changing gear	Clutch centre loose on gearbox mainshaft	Check for wear and retighten retaining nut.
Clutch noisy when withdrawn	Badly worn clutch centre bearing	Renew bearing.
Clutch neither frees nor engages smoothly	Burrs on edges of clutch plates and slots in clutch drums	Dress damaged parts with file if damage not too great.
Clutch action heavy	Dry operating cable or bends too tight	Lubricate cable and re-route as necessary.
Clutch action harsh	Overtight primary chain	Re-adjust primary chain.
Clutch 'bites' at extreme end of lever movement	Worn linings	Replace clutch friction plates.
Constant loss of clutch adjustment	Worn pushrod due to failure to maintain minimum clearance	Replace pushrod and re-adjust.

Chapter 4 Carburation and lubrication

Contents

Specifications

Carburettors

Make and type	Amal Concentric, left and right handed, type 930 (Type 932, Combat, late 750 cc, and all 850 cc)
Choke size	30 mm 32 mm (Combat, late 750 cc, and all 850 cc)

Carburettor settings

Model type	Standard Commando	Combat	850 cc
Main jet	220 *	220 †	260
Needle jet	0.107	0.106	928/104 #
Needle position	Middle **	Middle **	Top
Throttle valve	3	3	3½

* Varies on 1970 models, according to type of megaphone silencers fitted
† 210 if mute fitted
** Top, if mute fitted (1972 and Combat models)
# Must be used in conjunction with choke tube 928/107

Replaceable oil filter

Make	Crossland* AC*
Type	631* X4*

* or equivalent in another range

1 General description

The fuel system comprises a petrol tank astride the large diameter top frame tube from which petrol is fed by gravity to the twin Amal Concentric carburettors. Two petrol taps, each with a built-in gauze filter, are located at the lower rear end of the petrol tank. If only the left hand tap is used, the other can be turned on to provide a small quantity of petrol when the main body of the tank is empty, thus acting as a reserve.

A large capacity air cleaner with a replaceable corrugated paper element is connected to the intakes of both carburettors. It is located immediately to the rear of the carburettors, below the rear end of the petrol tank.

The twin Amal carburettors have their throttle slides and air slides linked together by means of separate junction boxes so that only one cable emerges from the twist grip and the air lever.

The air slides are used only for cold starting, when they act as a choke.

Two basic types of exhaust system are used, the more common twin downswept type using individual exhaust pipes and silencers. The 850 cc models have the two exhaust systems united by means of a balance pipe, located immediately below the initial bend from the exhaust ports. The alternative high-level exhaust system supplied with some models has separate exhaust pipes and silencers but is arranged so that both systems are carried close together on the left hand side of the machine, well above the primary chaincase.

Lubrication is effected on the dry sump principle, in which a gear-type mechanical pump feeds oil under pressure to the various engine components, via filters and a pressure release valve. Excess oil is pumped from the crankcase back to the oil tank.

2 Petrol tank - removal and replacement

1   Before the petrol tank is removed from the machine, both petrol taps should be closed and the petrol pipes detached by unscrewing the union joints. There is no necessity to drain the tank unless it is desired to remove either of the petrol taps.
2   Methods of tank mounting vary, according to the model and the type of petrol tank fitted. All locate with a short metal plate welded across the lower top frame tube, immediately to the rear of the steering head. It is necessary only to remove the locknuts and rubber insulating washers to free the nose of the tank.
3   At the rear of the petrol tank, a short strap unites both halves and is secured by means of studs which project from the underside of the tank. Alternatively, a rubber ring which passes under the frame tube is located with two 'hooks' at the rear of the tank. When either of these fixing methods is released, the tank can be lifted away from the frame. Note the location of the two rubber pads on top of the main frame tube which insulate the tank from vibration. They must be replaced in the same position.
4   Replacement is accomplished by reversing the procedure detailed in the preceding paragraphs. Check that the rubber ring is located fully with the 'hooks' if this rear fixing method is used, and that the vent hole in the filler cap is not obstructed. If the tank is airtight, the supply of petrol will be cut off, leading to a mysterious engine fade-out which is difficult to eliminate without realising the cause.

3 Petrol taps - removal and replacement

1   Both petrol taps thread into an insert in the bottom of the petrol tank, one on each side. The taps seat on a fibre washer which should be renewed to obviate leakage, each time the taps are removed and replaced.
2   The taps are of the lever type and are unlikely to give trouble during the normal service life of the machine. If the rate of flow becomes restricted, it is probable that the gauze filter within the petrol tank has become choked. Under these circumstances it will be necessary to drain the petrol tank and unscrew the defective tap so that the filter can be cleaned.

4 Petrol feed pipes - examination

1   The petrol feed pipes are fitted with unions to make a quickly detachable joint at both the carburettor float chamber and the two petrol taps. Leakage is unlikely to occur unless the union nuts slacken, the tubing splits or the metal ferrules around the pipe ends work loose.
2   After a long period of service, the transparent plastic material of which the pipes are made will harden and discolour due to the gradual removal of the plasticiser by the petrol. If the pipes are exceptionally rigid, they should be renewed because it is under this condition that they are most likely to crack,

especially in cold weather.
3   Never use ordinary rubber tubing, even as a temporary replacement. Petrol causes rubber to swell and disintegrate, thereby blocking the fuel supply completely.

5 Carburettors - removal

1   Commence by removing the petrol feed pipes at the banjo union joint with the underside of each float chamber. There is a nylon filter within each banjo union which will be displaced when the petrol feed pipe is withdrawn.
2   Detach the short rubber hose from each carburettor intake so that the air cleaner is disconnected. Remove the two crosshead screws in the top of each carburettor and lift the top away complete with control cables and the throttle valve and air slide assemblies.
3   Remove the carburettors as a pair by slackening and withdrawing the four socket screws which retain them (complete with curved distance pieces) to the cylinder head. Do not lose the heat insulating washers which will be displaced as the carburettors are withdrawn. The carburettors will be held together by the plastic balance pipe which interconnects both curved distance pieces. This can be removed when the carburettors are clear of the machine.
4   Tape the carburettor tops and slide assemblies to some nearby frame member to obviate the risk of damage when further dismantling occurs.
5   Note that the carburettors are identical in specification apart from the fact that they are 'handed'. This is necessary to ensure the pilot jet screw and the throttle stop screw are always outward facing to facilitate ease of adjustment.

6 Carburettors - dismantling

1   Both carburettors are virtually identical and in consequence the same dismantling procedure and layout of the component parts applies to each.
2   Commence by removing the float chamber. This is retained to the underside of the mixing chamber by two crosshead screws and can be lifted away when the screws are withdrawn. It will contain the horseshoe-shaped float, float needle and float spindle. Take care not to damage the gasket between the float chamber body and the base of the mixing chamber.
3   Drain the float chamber, then lift out the float, complete with float needle and spindle. Withdraw both the float needle and the spindle from the float assembly; both these components are small and easily lost. Place them in a safe place for subsequent examination.
4   The main jet can be seen projecting from the base of the mixing chamber. It should be unscrewed, followed by the jet holder above it which has the needle jet threaded into the furthermost end. Unscrew and detach the needle jet. Unscrew the pilot jet.
5   If the throttle stop screw and pilot jet screw are removed from the carburettor, note should be made of their settings so that they can be replaced in approximately the same position. Count the number of turns and part turns from the fully closed position.
6   Reverting to the carburettor tops which have been taped to a frame tube, lift the throttle valve return spring and disengage the spring clip from the needle, after making note of the notch with which the clip was engaged. Then disengage the end of the throttle cable from the throttle valve and lift the valve away.
7   The air slide assembly can be disconnected by disengaging the air cable nipple in similar fashion. However, it is unlikely that this assembly need be disturbed since the air slide rarely requires attention.
8   When dismantling the second carburettor, keep the component parts separate. Problems may occur if parts are unwittingly interchanged, especially moving parts subject to wear, such as the throttle valves.

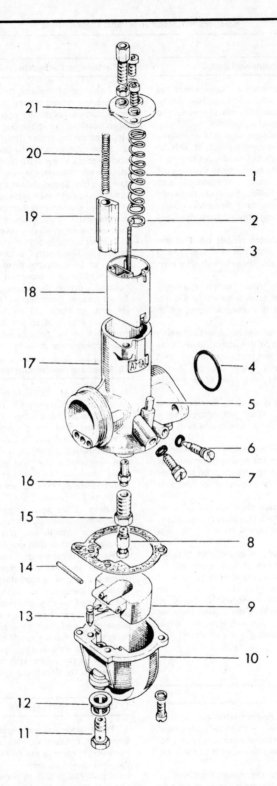

FIG. 4.1. COMPONENT PARTS OF THE AMAL CONCENTRIC CARBURETTOR

1 Throttle return spring	7 Throttle stop screw	13 Float needle	19 Air slide (choke)
2 Needle clip	8 Main jet	14 Float hinge	20 Air slide return spring
3 Needle	9 Float	15 Jet holder	21 Mixing chamber top
4 'O' ring	10 Float chamber	16 Needle jet	
5 Tickler	11 Banjo union bolt	17 Mixing chamber body	
6 Pilot jet screw	12 Filter	18 Throttle valve (slide)	

7 Carburettors - examination of the component parts

1 If the carburettor has shown a tendency to flood, check the float for leaks. Any leakage will immediately be obvious, due to the presence of petrol within the float. It is not practicable to repair a leaking or damaged float; renewal is essential.

2 The float needle and float needle seating should be examined with a magnifying glass. If wear has taken place, it will be evident in the form of a ridge on both the needle point and in the needle seating. This will prevent the needle from shutting off the fuel supply completely and will cause a permanently rich mixture, at the expense of fuel consumption. Renew both needle and seating if this form of wear is evident, or if the needle is bent or mis-shapen. The float needle seating threads into the float chamber body.

3 Check that the float needle has a Viton rubber tip. This late modification has been found advantageous in overcoming carburettor flooding problems, especially those initiated by high frequency vibration.

4 The float hinge is unlikely to give trouble unless it is badly worn or has been bent as the result of careless assembly. Do not attempt to straighten a bent hinge. It is a low cost item, which should be renewed if damaged.

5 Carburettor jets may block occasionally due to foreign matter which may have been present in the petrol. Never clear a blocked jet with a piece of wire or any sharp instrument since this will enlarge the jet orifice and cause changes in the carburation. Use either compressed air, or a blast of air from a tyre pump.

6 If petrol consumption has shown a tendency to rise, renew both the needle and the needle jet. Wear will occur after a lengthy period of service. Make sure the needle is not bent and that the clip is a good fit.

7 Wear of the throttle valve will be self-evident and may be accompanied by a clicking noise when the engine is running due to the valve rattling within the mixing chamber body. Wear marks are usually found at the base of the valve, on the side nearest the inlet port.

8 Check that neither the throttle stop screw nor the pilot jet screw is bent and that the taper of the latter screw is not worn. Check the threads for soundness and renew the small O rings which form a small but effective seal.

9 If the throttle valve has worn badly, it is probable that the mixing chamber has worn too. Evidence of such wear will be found in the vicinity of the throttle valve bore, close to the inlet and outlet passages. A worn carburettor body cannot be reclaimed; it must be renewed and a new throttle valve fitted.

10 Blow out all the internal air passages of the mixing chamber and check that the internal threads are sound, especially those that accept the crosshead screws retaining the carburettor top and the float chamber.

8 Reassembling the carburettors

1 Reassemble the carburettors by reversing the dismantling procedure. If the float chamber gasket is damaged, fit a new replacement and check that it is fitted the correct way round, so that the holes align with the jet passages and the ends of the float spindle are not trapped.

2 When refitting the carburettor top, the needle must engage with the needle jet and the air slide with the cutaway in the top of the throttle valve. Do not use force, it is quite unnecessary if everything locates correctly.

3 Be sure to fully tighten the crosshead screws which retain both the carburettor top and the float chamber. If the top works loose, the throttle may jam open, whilst if the float chamber is not secured rigidly, petrol spillage is inevitable.

4 Before reconnecting the petrol feed pipes, check that the nylon filter within the banjo union at the base of the float chamber is clean and not crushed.

9 Carburettors - checking the settings

1 The sizes of the jets, throttle valves, needles and needle jets are predetermined by the manufacturer and should not require modification unless the engine has been tuned. Check with the Specifications list at the beginning of this Chapter if there is any doubt about the values fitted.

2 Slow running is controlled by a combination of throttle stop and pilot jet screw settings. Commence by screwing the throttle stop screws inwards the same amount until the machine runs at a fast tick-over speed. Adjust the pilot jet screws until the tick-over is even, without either misfiring or hunting. Unscrew both throttle stop screws an identical amount until the desired tick-over speed is obtained, and again check both pilot jet air screws so that the tick-over is even. Always make these adjustments with the engine at normal working temperature, otherwise false settings may be obtained. The normal setting for the pilot jet screw is in the region of 1½ complete turns out from the fully closed position. The mixture is enriched by turning this screw INWARDS because it meters the supply of air and not fuel.

3 As a rough guide, up to 1/8 throttle is controlled by the pilot jet setting, from 1/8 to 1/4 throttle by the throttle valve cutaway, from 1/4 to 3/4 throttle by the needle position and from 3/4 to full throttle by the size of the main jet. These are only approximate divisions; there is a certain amount of overlap.

10 Balancing twin carburettors

1 On machines fitted with twin carburettors, maximum performance can be achieved only if both carburettors are completely in phase with one another. They must commence to open at the same time and must remain in complete sunchronisation throughout the entire throttle opening range.

2 Commence the check with a dead engine. Open the throttle slides by turning the twist grip and make sure that both throttle slides commence to rise at the same time. If they do not, adjust the individual throttle cable adjusters until the slides are completely in phase.

3 Check at the other end of the scale by opening the throttle slides fully by turning the twist grip. Neither should obstruct the carburettor intake and as they are lowered they should be completely in step with one another.

4 Remove the spark plug lead and spark plug from one cylinder and start the engine. Adjust the carburettor of the cylinder which is firing until a satisfactory tick-over speed is obtained, by following the procedure detailed in Section 9.2 of this Chapter. It is assumed the engine is already at normal running temperature before this adjustment is made. Stop the engine, replace the spark plug and lead, then repeat the procedure for the other cylinder. When both spark plugs are replaced and the engine is re-started, a satisfactory low speed tick-over should now be achieved. If the engine still runs a shade too fast, the speed can be lowered by slackening each throttle stop screw an identical amount.

11 Induction system joints

1 There is generally no necessity to detach the curved spacers when the carburettors are removed from the cylinder head. However, if for any reason the joint is broken, it is essential that the O ring in the centre of each carburettor flange is in good condition and is seating correctly in its groove. An air leak can cause a weak mixture, which may eventually result in a burnt piston or valves.

2 Do not overtighten the carburettor at the flange to spacer joint. Overtightening will cause the carburettor flange to bow and initiate an air leak which the O ring cannot seal. A bowed flange can be reclaimed by removing the O ring and rubbing the flange on a sheet of emery cloth wrapped around a sheet of plate glass, using a rotary motion. Check with a straight edge and do not omit to clean thoroughly before the O ring seal is replaced.

3 Do not overlook the heat insulators between the curved

spacers and the cylinder head joint. They break up the heat path and prevent the carburettors from overheating by conduction.

12 Air cleaner - dismantling, servicing and reassembling

1 The air cleaner chamber contains a corrugated paper element around its periphery which is protected by a perforated metal band. Access to the element is gained by removing the left hand side cover of the machine, detaching the rubber hoses from the carburettor intakes and slackening and removing the two bolts which retain the front plate. The plate can now be manipulated outwards from the bottom so that the element and protective metal band can be withdrawn together.

2 Unless heavily contaminated, wet or soaked in oil, the element can be re-used after it has been tapped on the bench to dislodge any foreign matter and then blown clear with an air line. Although the paper is resin-impregnated, it should be handled with care. If it is torn or perforated, renewal will be necessary.

3 Replace the element by reversing the dismantling procedure. Tighten the two retaining bolts after a check has been made to ensure neither the element nor the perforated metal band is trapped at any point.

4 It follows that the carburettor intake hoses should be in good condition, otherwise air leaks will occur. On no account run the machine with the air cleaner disconnected or without an element. The carburettors are jetted to take the presence of the air cleaner into account and if this advice is overlooked, a permanently weak mixture will result.

13 Exhaust system - general

1 Mention has been made of the two different types of exhaust system available which, irrespective of their appearance, form a separate complete system for each cylinder. The rubber suspension blocks at the silencer mounting points should be examined at frequent intervals to make sure they have not deteriorated.

2 Absolute tightness of the finned exhaust pipe nuts is essential. If they work loose, they will chatter away the internal threads within the exhaust ports, making it extremely difficult to devise any alternative method of attachment. Although more unsightly than the retaining lock washers, the finned nuts can be drilled and wired together so that if one nut commences to slacken, it will tend to tighten the other.

3 A copper-asbestos sealing washer must be located between the exhaust port opening and the end of the exhaust pipe. This will obviate the occurrence of air leaks, which often cause a mysterious and difficult to eliminate backfire which occurs only on the over-run.

4 Various designs of silencer are fitted, some of which conform to the noise reduction requirements of certain overseas countries. Unlike the exhaust system of a two-stroke, the silencers are unlikely to require cleaning out at regular intervals.

14 Lubrication system - general

The lubrication system functions on the dry sump principle. Oil from a separate side-mounted oil tank is fed by gravity and suction to the feed section of a gear-type pump housed in the right hand crankcase and driven from the right hand end of the crankshaft. The pump delivers oil under pressure through the crankshaft to both big ends via a pressure release valve. Oil escaping from the big end bearings lubricates the cylinder walls, main bearings and camshaft by splash, whilst a small bleed-off from the main supply is utilised to lubricate the rocker gear. The inlet rocker box drains via an oilway in the cylinder barrel and the exhaust rocker box through an oilway into the pushrod tunnel, where the return flow of oil lubricates the cam followers. Oil which collects in the crankcase sump is picked up by the scavenge section of the oil pump and returned to the oil tank for recirculation. A bleed-off from the return feed passes to a regulator and is used to lubricate the rear chain.

15 Dismantling, renovating and reassembling the oil pump

1 Unless the oil pump gives trouble, it should not be dismantled unnecessarily. Oil pump faults can be divided into two categories; complete failure of the pump and a tendency for oil to leak through the pump whilst the machine is stationary, causing the crankcase to fill with oil. The first type of fault is usually associated with some form of breakage in the oil pump drive, or the presence of metallic particles in the pump itself which has caused the gears to lock, shearing one of the drive spindles. The less serious oil leakage is caused by general wear and tear which can be rectified to an extent by taking up end float, as described in the following paragraph. Any fault attributed to the oil pump necessitates a complete strip down for examination.

2 Remove the oil pump as a complete unit from within the timing chest, as described in Chapter 1.11, paragraphs 1 and 2. The degree of end float can be ascertained by moving the oil pump driving gear in relation to the pump body; when the pump is new there should be a complete absence of end float.

3 To strip the pump, remove the four long screws which hold the oil pump together. The top cover can be lifted off together with the drive gear and spindle; the other pump gears can be lifted out, also the second spindle. The gears are keyed onto the spindles but are only a light push fit.

4 Wash all the components in neat petrol and allow them to dry. It will be noted that the scavenge section of the pump is twice the capacity of the feed section, necessary to keep the crankcase clear of excess oil. In consequence, the scavenge pump gears have wider teeth.

5 Whilst the pump is stripped out, take the feed pump end of the pump body (end housing the feed gears) and rub it down on a sheet of fine emery cloth wrapped around a perfectly flat surface, such as a sheet of plate glass. Rub down in easy stages, pausing at frequent intervals to wash off and reassemble the pump in order to check that the oil pump driving spindle will still revolve freely. Continue rubbing down until just the slightest amount of stiffness when turning the drive spindle is discernible. This shows that there is no longer excess clearance between the feed gears and the pump housing.

7 Repeat the process with the scavenge pump end of the pump body and check in similar fashion by assembling the pump with the scavenge gears only. Strip the pump again, wash all components with neat petrol, then reassemble and tighten the four screws fully. At this stage, some stiffness in the complete pump should be apparent.

8 Prime the pump with clean engine oil by pumping oil from a pressure oil can into the main feed orifice, whilst turning the driving spindle so that oil will circulate through the gears. The pump should now revolve more freely, the slight stiffness decreasing or even disappearing completely.

9 It follows that if metallic particles were found when the pump was stripped and if any of the gears have chipped or broken teeth, the pump should be renewed as a complete unit. Remember that if the oil pump does not function correctly, serious engine damage will result.

10 The oil pump driven gear is keyed onto the pump spindle and is retained by a self-locking nut. Replacement of the gear pinion is easy, especially since it is a parallel fit on the shaft.

16 Oil pressure release valve

An oil pressure release valve is fitted in the left hand end of the timing cover to prevent the oil pressure from rising above 45 to 55 psi. It comprises a spring-loaded plunger pre-set during manufacture of the machine by inserting shims so that it will actuate at the desired pressure. It requires no attention, other than the occasional check that the dome nut is tight and there is no oil leakage.

16.1 Pressure release valve is pre-set and requires occasional cleaning only

17 Crankcase breather

1 All models prior to 1972 have a timed and ported crankcase breather which takes the form of a spring-loaded rotary disc with cutaway segments, driven from the left hand end of the camshaft. The cutaways align with similar cutaways in a stationary disc behind the left hand camshaft bush at a predetermined time, which cannot be changed if the driving dogs of the rotary disc align correctly with the corresponding cutaways in the end of the camshaft.

2 Post 1972 750 cc models and the 850 cc models are fitted with a quite different non-mechanical breather. This is located at the rear of the left hand crankcase and comprises a micro-cellular foam separator, held by two retainer discs and a cap.

3 A breather pipe is attached to the outlet of either type of breather and is connected to the air filter backplate.

18 Oil filters - location and cleaning

1 All models have a gauze type filter incorporated in the main oil feed pipe from the oil tank. When the oil tank is drained for an oil change, this filter should be removed, washed in petrol and dried, before replacement. It forms an integral part of the nut of the oil pipe banjo union.

2 Pre-1972 and 850 cc models have a sump filter in the base of the left hand crankcase, identified by the very large bolt which forms the housing for the filter element. The bolt should be unscrewed and the gauze filter removed for cleaning with petrol by releasing the circlip and washer which precedes it.

3 All post-1972 models have a crankcase drain plug in the left hand crankcase, which is of the magnetic type. The drain plug should be removed during oil changes and any particles adhering to the magnet removed.

4 An additional, separate car-type oil filter is fitted to all post-1972 models. It is located between the rear engine plates, in the space between the rear of the gearbox and the rear wheel. The filter element is of the screw-on type and should be renewed during each oil change. It is retained by a large diameter hose clip around a locking plate on the right hand side of the engine plates, as a safeguard against the element unscrewing. In most cases, the element is an extremely tight fit and it will be necessary to use a strap spanner to free it initially. The element, which is replaceable and cannot be reconditioned, has a normal right hand thread.

19 Fault diagnosis

Symptom	Cause	Remedy
Engine 'fades' and eventually stops	Blocked air hole in filler cap	Clean.
Engine difficult to start	Carburettor flooding	Dismantle and clean carburettor. Check for punctured float.
Engine runs badly. Black smoke from exhausts	Carburettor(s) flooding	Dismantle and clean carburettor. Check for punctured float.
Engine difficult to start. Fires only occasionally and spits back through carburettors	Weak mixture	Check for fuel in float chambers and whether air slides working.
Oil does not return to oil tank	Damaged oil pump or oil pump drive	Stop engine immediately. Check oil tank contents. Remove timing cover to examine oil pump drive.
Engine joints leak oil badly	Pressure release valve inoperative	Dismantle valve and clean.
Crankcase floods with oil when machine is left standing	Worn oil pump	Dismantle oil pump and eliminate end float of gears.
Excess oil ejected from crankcase breather	Cylinders in need of rebore Broken or damaged piston rings	Rebore cylinders and fit oversize pistons. Dismantle engine and replace rings.

Chapter 5   Ignition system

Contents

Specifications

Ignition coils

Manufacturer	Lucas
Type	17M6 or 17M 12 †
Voltage	6 volts

Contact breaker

Manufacturer	Lucas
Type	6CA or 10CA (late models)
Gap	0.015 in (0.35 - 0.4 mm) fully open

Capacitor

Manufacturer	Lucas
Type	2MC (Electrolytic)
Voltage	12 volts

Ballast resistor

Manufacturer	Lucas
Type	3BR

Spark plugs

	Champion	NGK
Manufacturer		
Type	N7Y *	BP7ES
Size	14 mm	
Reach	¾ inch	
Gap	0.023 - 0.028 in (0.59 - 0.72 mm)	

    † 17M12 coils fitted to some 1972 models. No ballast resistor required
    * Originally N6Y

1 General description

The spark necessary to ignite the petrol/air mixture in the combustion chambers is derived from a battery and twin ignition coils. Each cylinder has its own separate contact breaker, condenser and coil to determine the precise moment at which the spark will occur. When the contact breaker points separate, the low tension circuit is broken and a high tension voltage is developed by the coil, which jumps the air gap across the points of the spark plug and ignites the mixture.

An alternator connected to the left hand end of the crankshaft assembly generates an alternating current which is rectified and used to charge the 12 volt, ten amp/hr battery. The twin ignition coils are suspended horizontally from the petrol tank front mounting, together with their associated condensers. The twin contact breaker assembly is contained within the circular housing at the top right hand side of the timing cover, or in the case of the very early models, within a housing attached to the rear of the timing chest. The battery is located behind the left hand side panel, beneath the seat. The circuit is actuated by an ignition switch mounted on the left hand rear frame tube, in line with the carburettors.

2 Crankshaft alternator - checking the output

Unlike a great many other modern machines, the early Norton Commando has an ammeter mounted in the top of the headlamp shell which gives an immediate indication of the output from the alternator whilst the engine is running. If no charge is indicated, even with a full lighting load, it is probable that the alternator is at fault. This type of problem is covered at greater length in Chapter 8 under the Electrical System heading, which should also be consulted if the machine has no ammeter fitted and the red charge lamp does not extinguish.

3 Ignition coils - checking

1 Each ignition coil is a sealed unit, designed to give long service without attention. The coils are mounted close together, suspended horizontally from the petrol tank front mounting. If a weak spark and difficult starting cause either coil to be suspect, it should be checked by a Norton Villiers agent or an auto-electrical expert who will have the appropriate test equipment. A faulty coil must be renewed; it is not possible to effect a satisfactory repair.
2 It is extremely unlikely that both coils will fail simultaneously. If the complete ignition circuit fails, it is highly probable that the source of the fault will be found elsewhere. Apart from the common low tension supply, both coils work on independent circuits.
3 A failed condenser or a dirty contact breaker assembly can give the impression of a faulty ignition coil and these components should be checked first before the coil itself receives attention.

4 Contact breakers - adjustments

1 To gain access to the contact breaker assembly, remove the circular, chromium-plated contact breaker cover which is retained to the top right hand corner of the timing cover by two screws, or on early models, the wire clip which holds the cover in position. (Unit mounted on rear of timing chest.)
2 Rotate the engine slowly by means of the kickstarter until one set of contact breaker points is in the fully open position. Examine the faces of the contacts; if they are pitted or burnt, it will be necessary to remove them for further attention, as described in Section 5 of this Chapter. Repeat this action for the other set of points.
3 Adjustment of the contact breaker gap is achieved by releasing the screw which retains the fixed contact breaker point in position (screw D in the accompanying illustration) and rotating the eccentric screw C until the correct gap of 0.015 inch is obtained. Screw D should then be tightened, to lock the points in position. Always re-check before passing to the second set of points which are adjusted in similar fashion.
4 It is important to ensure that the gap is re-set when the points are fully open, otherwise a false reading will result. It is equally important that both sets of points have the same gap.
5 Before replacing the contact breaker cover, smear the contact breaker cam sparingly with grease and add a few drops of engine oil to each lubricating felt. The cover plate must be replaced with the small drain/breather hole facing downwards, to prevent the ingress of water.

5 Contact breaker points - removal, renovation and replacement

1 If the contact breaker points are pitted or burned, they should be removed for dressing. Badly worn points should be renewed without question, as should points which will require a substantial amount of material to be removed before their faces can be restored.
2 To remove the contact breaker points, remove the nut securing the spring to the terminal post and the small circlip which retains the rocker arm of the contact on its pivot. Lift the rocker arm and contact away, making special note of the way in which the insulators are arranged. If these are reassembled in incorrect order, the moving contact will be earthed and the contact breaker isolated electrically.
3 Removal of the locking screw will free the fixed contact

plate.
4 The points should be dressed with an oilstone or fine emery cloth. Keep them absolutely square during the dressing operation, otherwise they will make angular contact with each other when reassembled and will quickly burn away.
5 Clean the points with petrol to remove all traces of abrasive and oil, then reassemble them by reversing the dismantling procedure. Make sure the insulating washers and collar are replaced in the correct order and that the pivot of the moving contact rocker arm has a light smearing of grease.
6 When reassembly is complete, check and if necessary, re-set both contact breaker gaps.

6 Auto-advance assembly - examination

1 The auto-advance assembly is located behind the contact breaker baseplate and it is necessary to remove the baseplate as a complete unit to gain access. This is accomplished by unscrewing the two screws which retain the contact breaker baseplate assembly through elongated slots.
2 The auto-advance assembly comprises two spring-loaded weights which fly apart with centrifugal action. They are connected to the contact breaker cam so that as the engine speed rises, the cam moves independently of the camshaft from which the assembly is driven, thereby advancing the ignition timing.
3 Visual inspection will show whether tired or broken springs are restricting the freedom of movement, or whether rust, resulting from condensation, is preventing the weights from moving smoothly. The only maintenance required is an occasional drop of oil on the various pivots to ensure they do not run dry. Check that the weights return to the closed position when the contact breaker cam is turned anticlockwise and then released.

7 Condensers - removal and testing

1 The condensers are included in the contact breaker circuitry to prevent arcing across the contact breaker points at the moment of separation. Each condenser is connected in parallel with its own set of points and if the condenser should fail, an ignition fault will occur in the circuit involved. For convenience, the condensers are mounted together in the form of a pack on all models from 1971 onwards.
2 If the engine is difficult to start or if a persistent misfire occurs on one cylinder, it is possible that the condenser in the ignition circuit of that cylinder has failed. To check, separate the contact breaker points of the cylinder concerned by hand, with the ignition switched on. If a spark occurs across the points and they have a blackened and burnt appearance, the condenser can be regarded as unserviceable.
3 It is not possible to test the condenser without the appropriate test equipment. It is best to fit a new replacement in view of the low cost involved and observe the effect on engine performance.
4 Post 1971 models have the condenser pack attached to the coil cluster. Earlier models have the condensers attached individually to each coil clip. They can be removed and replaced without need to disturb the coils.

8 Ballast resistor - location and testing

1 All models fitted with 6 volt coils (some 1971 models and all later models) require what is known as a ballast resistor to permit their use in an otherwise 12 volt system. The advantage of using 6 volt coils is the ability to produce a high intensity spark when the battery is in a discharged condition. The ballast resistor permits the coils to work at their approximate potential since it is by-passed when the voltage is low, permitting the coils to be fed direct.
2 Although the ballast resistor is unlikely to give trouble, it can be checked occasionally after the electrical leads have been

temporarily detached, using a multimeter. The ballast resistor has a value of 1.8 - 2 ohms. It is located with the ignition coil assembly.

9 Ignition timing - checking and resetting

1 If the ignition timing is approximately correct, the line inscribed on the alternator rotor will coincide exactly with the 28° mark on the indicator plate within the primary chaincase, when the contact breaker points of the cylinder being checked are on the point of separation. The scale is exposed by removing the circular threaded cap in the top left hand corner of the chaincase. This check must be made with the contact breaker locked in the fully advanced position. Norton Villiers special washer 060949 is supplied for this purpose. It fits under the auto-advance cam centre bolt and will lock the cam in the fully-advanced position if the cam is held in this position whilst the centre bolt is tightened. A washer with a hole large enough to clear the centre unit and press on the cam when tightened down can be used as a substitute, with equally good effect. The cylinder being checked or re-timed must be on the compression stroke and the engine should be turned forwards when arriving at this setting, to eliminate any backlash.
2 The factory arrangement for timing is for the left hand cylinder to be timed in conjunction with the contact breaker to which the black and yellow electrical lead is attached (the left hand set of points, viewed from the contact breaker end). It is imperative that the contact breaker points are correctly gapped at 0.015 inch before making this check, since any resetting afterwards will affect the accuracy of the ignition timing.
3 No problem should occur in obtaining an accurate setting for both cylinders, since the contact breaker points can be moved independently of each other, to either advance or retard the setting until the correct reading is obtained. This involves slackening screws A as shown in Fig. 5.1 and using the eccentric screws B to either advance or retard the points assembly. Do not omit to lock screws A after the timing has been checked and agreed accurate, then check again to ensure the points did not move as the screws were tightened.
5 If it is difficult to ascertain when the points separate, connect one lead of a voltmeter to the spring of the moving contact point and the other lead to earth. Switch on the ignition; the voltmeter will show a reading immediately the points separate. If a voltmeter is not available, use a strip of cigarette paper between the points. If a pull is maintained on the paper, the grip on the paper will be released as the points separate.
6 For optimum performance, it is preferable to confirm the accuracy of the ignition by the use of a stroboscope, an electrical device which will not be available to most owners. A stroboscope check can, however, be arranged with most Norton Villiers agents; it is easy to effect since the check is carried out with the engine running and virtually no dismantling.

10 Ignition timing - checking by stroboscope

1 Remove the screwed inspection cap in the top left hand corner of the primary chaincase, to expose the degree indicator plate and timing mark on the alternator rotor.
2 Remove the circular cover from the top right hand corner of the timing cover, to give access to the contact breaker assembly.
3 Connect the stroboscope to the right hand cylinder plug lead and use a separate battery to power the stroboscope lamp. If the battery on the machine is used, feed back into the stroboscope circuit can occur and give a false reading.
4 Identify the set of contact breaker points on which the timing is to be set.
5 Start the engine and run at a steady 3000 rpm as indicated on the tachometer. Shine the stroboscope lamp on the indicator plate and check the reading. If the timing mark corresponds exactly with the 28° line on the indicator plate, the right hand cylinder is timed correctly. Repeat the procedure for the left

hand cylinder, connecting the stroboscope into the left hand spark plug lead.
6 If the timing mark does not correspond with the 28° line, the contact breaker points associated with the cylinder involved should be moved clockwise to advance the timing or anti-clockwise to retard it. This is facilitated by slackening screw A (Fig. 5.1.) and using eccentric screw B to move the points in relation to the contact breaker cam. Tighten screw A and recheck with the stroboscope. Continue until the timing mark corresponds exactly for both cylinders.
7 Disconnect the stroboscope and replace both the contact breaker end cover and the inspection cap in the primary chaincase.

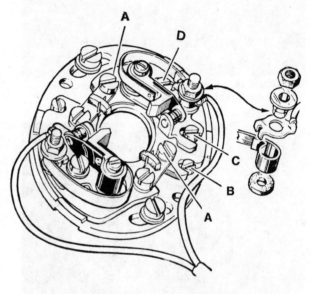

Fig. 5.1. Lucas contact breaker

11 Capacitor - function, location and testing

1 An electrolytic capacitor encased within a spring is mounted to the rear of the battery. The capacitor is fitted to provide an emergency start facility so that the machine can be started even when the battery is fully discharged by utilising a system of energy transfer from the alternator.
2 Whilst a defunct capacitor will not prevent the ignition circuit from working whilst the battery is in a state of charge, it will eliminate the emergency start facility, should the occasion arise. In consequence it is advisable to check the capacitor periodically.
3 To test the capacitor, remove the battery terminals and insulate the negative terminal to obviate risk of a short circuit. Start and run the engine, then switch on the full lighting load. If all these facilities are available without the battery connected, the capacitor is in good working order.
4 If the machine will not start, remove the capacitor and connect it to a 12 volt battery for about 5 seconds. It is most important that the polarity is correct; the negative lead from the battery must be connected to the double terminal. The small 3/16 inch terminal marked with a red spot on the rivet is the positive (earth) connection. IF THESE PRECAUTIONS ARE NOT OBSERVED, THE CAPACITOR WILL BE DESTROYED.
5 After the battery has been disconnected, connect a DC voltmeter across the capacitor terminals, observing the polarity. A reading of not less than 9 volts indicates a serviceable capacitor, ignoring any initial overswing of the meter needle. If the reading is less than 9 volts, the capacitor is leaking (electrically) and should be replaced.

9.1 Indicator plate in chaincase gives ignition advance when rotor timing mark aligns

9.1a Rotate points cam fully anti-clockwise and secure in place using a large washer as improvised locking tool (see text)

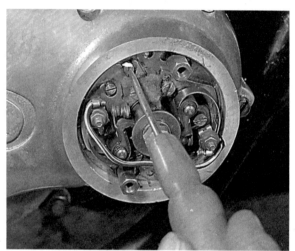

9.3 Eccentric screws facilitate close adjustment of points

10.5 Strobe lamp aligns timing mark on scale as though engine was static

12 Spark plugs - checking and resetting the gaps

1   14 mm spark plugs with a ¾ inch reach are fitted to all Norton Commando models. Refer to the Specifications section of this Chapter for the list of recommended grades. Always use the grade recommended, or the direct equivalent in another manufacturer's range.

2   Check the gap at the plug points every 2000 miles. To reset the gap, bend the outer electrode closer to the central electrode and check that a 0.023 inch feeler gauge can be inserted. Never bend the central electrode or the insulator will crack, causing engine damage if particles fall in whilst the engine is running.

3   The condition of the spark plug electrodes and insulator can be used as a reliable guide to engine operating conditions, with some experience. See accompanying illustrations.

4   Always carry at least one spare spark plug of the correct grade. This will serve as a get-you-home means if one of the spark plugs in the machine should fail.

5   Never overtighten a spark plug, otherwise there is risk of stripping the threads from the cylinder head, especially in the case of one cast in light alloy. A stripped thread can be repaired by using what is known as a 'Helicoil' thread insert, a low cost service of cylinder head reclamation that is operated by many dealers.

6   Use a spark plug spanner that is a good fit, otherwise the spanner may slip and break the insulator. The plug should be tightened sufficiently to seat firmly on its sealing washer.

7   Make sure the plug insulating caps are a good fit and free from cracks. The caps contain the suppressors which eliminate radio and TV interference; in rare cases the suppressors have developed a very high resistance as they have aged, cutting down the spark intensity and giving rise to ignition problems.

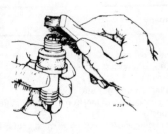

FIG. 5.2. SPARKING PLUG MAINTENANCE

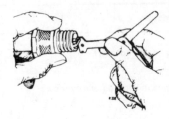

Cleaning deposits from electrodes and surrounding area using a fine wire brush

Checking plug gap with feeler gauges

Altering the plug gap. Note use of correct tool

White deposits and damaged porcelain insulation indicating overheating

Broken porcelain insulation due to bent central electrode

Electrodes burnt away due to wrong heat value or chronic pre-ignition (pinking)

Excessive black deposits caused by over-rich mixture or wrong heat value

Mild white deposits and electrode burnt indicating too weak a fuel mixture

Plug in sound condition with light greyish brown deposits

Sparking plug electrode conditions

13 Fault diagnosis

Symptom	Cause	Remedy
Engine will not start	Break or short circuit in electrical system Contact breaker points closed	Switch off and check wiring. Re-adjust contact breaker points.
Engine misfires on one cylinder	Faulty condenser Incorrect ignition timing Fouled or incorrect grade of spark plug	Replace condenser and re-check. Check accuracy of ignition timing of cylinder involved. Clean plug and/or replace with correct grade.
Engine lacks response and overheats	Reduced contact breaker gaps Jammed auto-advance unit	Check and reset. Check whether balance weights are free to move.
Engine 'fades' when under heavy load	Pre-ignition	Replace plugs, using only recommended grades.
Engine will not start with discharged battery	Faiied capacitor in emergency start system	Check capacitor and replace if necessary.

Chapter 6 Frame and Forks

Contents

Specifications

Forks

Type	Telescopic with two-way hydraulic damping. Internal springs
Total movement	6 in (15.24 cm)
Stanchion tubes	1.3590 - 1.3575 in (34.29 - 34.467 mm) outside diameter
Upper bush	1.3595 - 1.3605 in (34.53 - 34.48 mm) inside diameter, fitted
Lower bush	1.4980 - 1.4990 in (38.05 - 38.075 mm) outside diameter, fitted
Lower fork leg	1.4995 - 1.5010 in (38.087 - 38.125 mm) internal diameter

Fork springs

No of coils	75½ approx
Free length	18.687 in (474.65 mm)
Rating	36.5 lb in Marked with red paint

Fork leg capacity
Fork leg capacity ... ... 150 cc (5 fl oz) each leg

Engine mountings

Free play	0.010 in (0.254 mm)
Shim sizes available	0.005 in, 0.010 in, 0.020 in and 0.030 in (0.127 mm, 0.254 mm, 0.508 mm and 0.762 mm)

Rear suspension units

Manufacturer	Girling
Spring fitted length	8.4 in (213.36 mm)
Spring colour code	Red/yellow/red
Spring rating	126 lb in

1 General description

The patented Isolastic method of vibration damping employed on all Norton Commando models couples together the engine, transmission, swinging arm fork and rear wheel as a complete unit which can be removed in its entirety from the main frame assembly, if so desired. Isolation is achieved by the use of resilient mountings, as shown in the accompanying diagram. The engine unit oscillates on the rear mounting (B) which comprises three bonded and two rubber buffers. Because the swinging arm fork is mounted on the rear engine plates, it is isolated from the main frame assembly, thus obviating any misalignment that would otherwise occur between the engine and rear wheel sprockets when they are under load. This arrangement gives maximum support, whilst effectively isolating the engine unit from the frame. The front mounting (C) controls the degree of movement of the engine unit. The two bonded and two buffer rubbers permit a greater degree of flexibility than that of the rear mounting. The cylinder head steady (A) completes the triangulated mounting arrangement by controlling the lateral movement of the engine unit in the frame.

The degree of side play in the front and rear mountings must be kept within closely controlled limits, to obviate passing engine and transmission vibrations to the rider. PTFE thrust washers and shims are used to control this movement so that it is kept within design limits, without need for frequent attention.

The slimline front fork fitted to the Norton Commando is a development of the original Norton Roadholder fork used on some of the early models. Unspring weight has been reduced to a minimum and suspension improved by the adoption of two-way damping. A further innovation, adopted during 1971, is the use of non-adjustable ball journal bearings for the steering head assembly.

2 Front forks - removal from frame

1 It is unlikely that the front forks will need to be removed from the frame as a complete unit, unless the steering head bearings require attention or the forks are damaged in an accident. In the case of models fitted with a disc brake, it will be necessary to remove the complete hydraulic assembly as a preliminary. It is a sealed system and can be removed in this fashion, thus obviating the need to drain, refill and bleed to eliminate air bubbles.

2 To remove the hydraulic system, release the hydraulic pipe bracket from the right hand lower fork leg, close to the mudguard bridge. Remove the two bolts and washers which secure the brake caliper to the lug on the fork leg and lift the caliper away, with the hydraulic hose and brake pads still attached. At this stage, a spacer should be inserted between the two pads, to prevent them from being ejected from the caliper if the brake lever is operated accidentally. Remove the four crosshead screws which secure the master cylinder to the switch cluster on the right hand end of the handlebars. When the screws are withdrawn, hold the master cylinder but allow the switch cluster to drop and hang suspended by its leads. Disconnect both electrical connections from the switch on the master cylinder, after pulling back the protective plastics cover. Release the spring clips which retain the hydraulic hose to the right hand fork leg and lift the complete hydraulic system away. It can be laid down, away from the working area, until it is required for reassembly.

3 Position the machine firmly on a wooden box or block placed under the lower frame tubes and remove the front wheel. Since the hydraulic system has already been removed, the wheel can be freed by removing the nut at the extreme right hand end of the fork spindle and withdrawing the spindle by means of a tommy bar through the left hand end, whilst lifting the front wheel clear of the ground. The wheel bearing dust covers will be displaced as the wheel spindle is withdrawn and should be placed aside until reassembly commences. The wheel can now be withdrawn from the forks completely.

4 If the front wheel has a drum brake, a slightly different

procedure is required. Commence by disconnecting the front brake cable at the brake operating arm. It is secured to the arm by a spring clip and clevis pin. Slacken the pinch bolt in the end of the left hand fork leg and remove the nut from the right hand end of the spindle. Insert a tommy bar through the left hand end of the spindle so that the spindle can be withdrawn whilst the front wheel is supported by hand. The wheel can now be withdrawn from the forks, after the brake plate anchor has disengaged from the abutment on the lower right hand fork leg.

5 Unscrew and raise the large chromium plated bolts at the top of each fork leg. Slacken the nut at the base of each bolt, so that the fork damper rod can be unscrewed and the bolt freed. Lift off the speedometer and tachometer heads and allow them to hang suspended from their cables. Remove the two bolts securing the headlamp and leave the headlamp suspended.

6 On pre-1971 models, remove the chromium plated dome nut in the centre of the fork top yoke, together with its washer. The top yoke can then be driven off the steering head stem and stanchion tapers by hitting the underside of the yoke with a rawhide mallet. Remove the nut and dust cover from the stem and then drive the stem downwards through the steering head bearings, to release the forks as a complete unit, complete with mudguard. The latter is detached by removing the two nuts on the inside of each fork leg and the bolts which anchor the stays to the lower end of each fork leg.

7 1971 and later models require a modified technique because the steering head stem is attached to the top yoke. Commence by bending back the tab washer and unscrewing the nut at the bottom of the fork stem, below the bottom fork yoke. Drive the lower yoke downwards to clear the end of the stem, using a rawhide mallet. Collect the two bottom and one top O ring released by each headlamp bracket and top cover which will be freed in conjunction with the brackets and top covers as the complete fork assembly is withdrawn from the frame. The upper yoke, complete with head stem, is released from the steering head bearings by driving it upwards with a rawhide mallet. The dust cover and washer will be displaced during this operation.

8 If it is desired to remove the individual fork legs, without dismantling the complete fork assembly, replace the chromium plated bolts in the top of each fork leg so that at least six threads engage. With a block of wood interposed between the bolt head and the hammer, drive the bolt downwards in order to break the taper joint of the fork stanchions. If the pinch bolt in the lower fork yoke is slackened, the fork leg can then be withdrawn downwards until it is free of the fork yokes.

9 The Roadholder forks fitted to the very early models are virtually identical in construction to the pre-1971 Slimline forks but have additional outer shroud tubes.

3 Front forks - dismantling the fork legs

1 Drain the fork leg by inverting it over a suitable receptacle or by removing the drain plug just above the wheel spindle orifice in the lower fork leg. Unscrew the locknut at the upper end of the damper rod and withdraw the locating bush and fork spring.

2 Clamp the lower fork leg in a vice, using aluminium clamps to prevent marking the outer surface. The fork leg should be clamped in the horizontal position.

3 Slide the plastic gaiter up the fork tube and unscrew the threaded collar over which the gaiter normally seats. The collar, which has a right hand thread, may be tight, necessitating the application of a strap spanner. When the collar is unscrewed fully, grasp the fork stanchion with both hands and with a number of sharp upward pulls, displace the oil seal, paper washer and top bush. The stanchion can now be lifted away from the lower fork leg completely.

4 Remove the damper tube anchor bolt which is recessed into the end of the lower fork leg, within the portion that carries the wheel spindle. Do not lose the fibre washer on which the bolt seats. The damper rod with damper unit attached can then be lifted out of the lower fork leg as a unit. Secure the damper tube carefully but firmly in a vice (it is easily crushed) and

unscrew the alloy damper tube cap. This will free the damper rod. There is no necessity to dismantle the damper rod assembly further, unless the damper valve is to be renewed. In this instance, the damper rod should be clamped firmly in a vice so that the damper rod locknut and square washer can be removed to free the damper valve.

5   To gain access to the lower fork bush, remove the square section circlip from the bottom of the fork stanchion.

4 Front forks - general examination

1   Apart from the oil seals and bushes, it is unlikely that the forks will require any additional attention, unless the forks springs have weakened or the damper units have lost their efficiency. If the fork legs or yokes have been damaged in an accident, it is preferable to renew them. Repairs are seldom practicable without the appropriate repair equipment and jigs. Furthermore, there is also risk of fatigue failure.

2   Visual examination will show whether either the fork legs or the yokes are bent or distorted. The best check for the fork stanchions is to remove the fork bushes, as described in paragraphs 3 and 5 of the preceding Section, and roll the stanchions on a sheet of plate glass. Any deviation from parallel will immediately be obvious.

5 Front forks - renewal of oil seals

1   If the fork legs have shown a tendency to leak oil or if there is any other reason to suspect the efficiency of the oil seals, they should be renewed without question.

2   The oil seals are retained in the cupped portion of the lower fork legs by the screwed collar. They are displaced by following the procedure detailed in Section 3.3 of this Chapter. The new oil seal should be tapped into position with a tubular drift of the correct size, after the top fork bush and washer have been located in the lower fork leg first.

6 Front forks - examination and renewal of bushes

1   Some indication of the amount of wear in the fork bushes can be gained when the forks are being dismantled. Pull each fork stanchion out of the lower fork leg until it reaches the limit of its extension and check the side play. In this position the two fork bushes are closest together, which will show the amount of play to its maximum. Only a small amount of play which is just perceptible can be tolerated. If the play is greater than this, the bushes are due for replacement.

2   It is possible to check for play in the bushes whilst the forks are still attached to the machine. If the front wheel is gripped between the knees and the handlebars rocked to and fro, the amount of wear will be magnified by the leverage at the handlebar ends. Cross-check by applying the front brake and pushing and pulling the rear wheel backwards and forwards. It is important not to confuse any play which is evident with slackness in the steering head bearings, which should be taken up first.

3   The fork bushes can be removed from the stanchion when the circlip at the lower end of the stanchion has been detached.

7 Steering head bearings - examination and renewal

1   Before commencing to reassemble the forks, the steering head bearings should be inspected. Pre-1971 models have ball journal bearings which can be adjusted, whereas later models have bearings of the non-adjustable type. In either case the bearings should be examined to ensure there is no excessive play or roughness, which will necessitate renewal.

2   The bearings are a drive fit into the steering head assembly and should be driven out with a soft metal drift. Great care is necessary when displacing or refitting the steering head bearings,

2.4 Pinch bolt secures wheel spindle in drum brake models

2.5 Bolts are attached to damper rods within fork legs

2.6 Mudguard is connected to forks in three positions

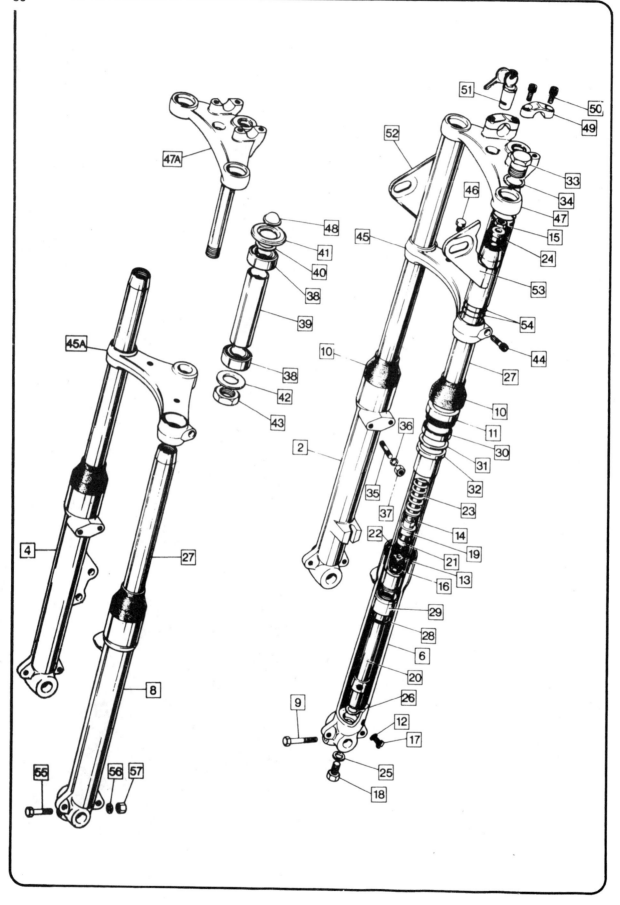

FIG. 6.1. FRONT FORKS

1 Right hand fork leg assembly, drum brake models
2 Right hand lower fork leg, drum brake models
3 Right hand fork leg assembly, disc brake models
4 Right hand lower fork leg, disc brake models
5 Left hand fork leg assembly, drum brake models
6 Left hand lower fork leg, drum brake models
7 Left hand fork leg assembly, disc brake models
8 Left hand lower fork leg, disc brake models
9 Pinch bolt (drum brake models only)
10 Gaiter 2 off
11 Screwed collar 2 off
12 Drain plug washer 2 off
13 Damper rod seat 2 off
14 Damper rod 2 off
15 Damper rod cap nut 2 off
16 Damper rod locknut 2 off
17 Drain plug 2 off
18 Damper tube anchor bolt 2 off
19 Damper tube cap 2 off
20 Damper tube 2 off
21 Damper valve 2 off
22 Damper valve stop pin 2 off
23 Fork spring 2 off
24 Fork spring locating bush 2 off
25 Damper tube anchor bolt washer 2 off
26 Damper tube anchor bolt washer 2 off
27 Stanchion 2 off
28 Stanchion bush circlip 2 off
29 Stanchion bush 2 off
30 Oil seal 2 off
31 Oil seal washer 2 off
32 Stanchion guide bush 2 off
33 Stanchion top bolt 2 off
34 Washer 2 off
35 Stud for mudguard bridge 4 off
36 Washer 4 off
37 Nut 4 off
38 Steering head bearing 2 off
39 Bearing spacer
40 Washer
41 Dust cover
42 Washer
43 Nut
44 Socket screw 2 off
45 Lower fork yoke - 750 cc models only
45a Lower fork yoke - 850 cc models only
46 Plug for lower yoke 2 off
47 Upper fork yoke with stem - 750 cc models only
47a Upper fork yoke with stem - 850 cc models only
48 Stem blanking cap
49 Handlebar clamp 2 off
50 Socket screw 4 off
51 Steering lock with keys
52 Right hand headlamp bracket and shroud
53 Left hand headlamp bracket and shroud
54 'O' ring 6 off
55 Bolt
56 Washer
57 Nut

2.8 Slacken pinch bolts in lower fork yoke to free individual fork legs

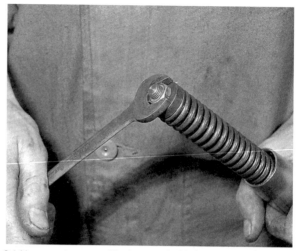

3.1 Unscrew damper rod locknut and withdraw bush and fork spring

3.3 When collar is unscrewed, fork stanchion is released from lower leg

3.4 Damper tube anchor bolt is recessed into end of lower fork leg

to prevent damage to the bearing tracks.

8 Front forks - reassembly

1   To reassemble the forks, follow the dismantling procedure in reverse. Particular care is necessary when refitting the oil seals or new replacements. Grease both the inside of the seal and the fork stanchion to obviate risk of damage to the feather edge of the seal.

2   Tighten the steering head carefully, so that all play is eliminated without undue stress on the bearings. The adjustment is correct if all play is eliminated and the handlebars will swing to full lock on their own accord, when given a light push on one end.

3   It is possible to place several tons pressure quite unwittingly on steering head bearings of the adjustable type, if they are overtightened. The symptom of overtight bearings is a tendency for the machine to roll at low speeds or the handlebars to oscillate, even though the handlebars may appear to turn quite freely when the machine is on the centre stand.

4   Difficulty may be encountered due to the reluctance of the stanchions to seat correctly on their tapers within the top fork yoke. The chromium plated bolts can be used to pull the stanchions into position, if necessary, without passing through the eye of the instrument case bracket as a temporary expedient.

3.4a Alloy cap secures damper rod to damper valve assembly

Do not use force unless a good depth of thread has engaged with the stanchion.

5 1971 and later models will require the use of a block of wood interposed between the upper fork yoke and a hammer in order to drive the head stem through the steering head bearings. They are a tight fit. Note that the washer must fit BELOW the top dust cover.

6 Tighten the chromium plated top bolts first, then the steering head stem nut, followed by the pinch bolts through the lower fork yoke. If the forks are incorrectly aligned or stiff in action, slacken each of them a little and bounce the forks several times before retightening from the top downwards, finishing with the pinch bolts. This same technique can be used if the forks are misaligned after an accident. Often the legs will twist within the fork yokes, giving the impression of more serious damage, even though none has occurred.

7 Do not omit to slacken and raise the two chromium plated fork bolts so that 150 cc of damping fluid can be added to each fork leg. Check that the drain plugs in each fork leg have been replaced first! The bolts should be tightened to a torque setting of 40 ft lb and the steering head stem nut to 15 ft lb (1971 and later models ONLY). The nut is then secured by a tab washer.

9 Front forks - damping action

1 Each fork leg contains a pre-determined quantity of damping fluid, used to control the action of the compression springs within the forks when various road shocks are encountered. If the damping fluid is absent, or the damper units ineffective, there is no control over the rebound action of the fork springs and fork movement would become excessive, giving a very 'lively' ride. Damping action restricts fork movement and is progressive in action; the effect becomes more pronounced as the rate of deflection increases. The Norton Villiers fork is particularly good in this respect, since the two-way damping system applies to both compression and extension.

2 Reference to the accompanying illustration will show how the Norton Villiers system operates. When the forks are compressed, the stanchion tube and damper rod remain stationary as the lower fork leg and damper tube rises. As the damper tube rises, the valve of the end of the damper rod is raised, permitting oil to pass the seating washer. Because the displaced oil cannot escape from the sealed end of the damper tube, excess oil is expelled through the bleed holes in the lower end of the fork stanchion, aided by the vacuum which is produced as the gap between the upper and lower fork bushes widens. It is the restrictive action of the bleeder holes which slows down the fork movement. When the lower fork leg approaches its limit of travel, a cone-shaped plug in the bottom of the leg commences to enter the end of the fork stanchion tube, thereby progressively cutting off the alternative passageway for the oil. Oil is therefore compressed in the extreme bottom end of the lower fork leg to provide an effective bump stop, leaving only the bleed holes in the damper tube as a controlled exit to prevent a complete hydraulic lock. When the fork extends, the lower fork leg and the damper unit descend, leaving oil trapped between the upper and lower fork bushes. Because the damper rod is also descending, the damper valve is forced against its seating and is held closed. The pressure exerted allows oil to escape via the tiny clearance between the damper rod and the cap of the damper unit, into the area above. As the fork continues to extend, the oil trapped between the fork bushes is compressed as the gap between the fork bushes closed and is forced into the lower fork leg through a large hole in the stanchion and then a smaller hole in the damper body, which effectively slows down fork action in a progressive manner and prevents the forks from reaching the upper limit of travel.

10 Frame assembly - examination and renovation

1 If the machine is stripped for an overhaul, this affords an excellent opportunity to inspect the frame for signs of cracks or other damage which may have occurred during service. Frame repairs are best entrusted to a frame repair specialist who will have all the necessary jigs and mandrels necessary to ensure correct alignment. This type of approach is recommended for minor repairs. If the machine has been damaged badly as the result of an accident and the frame is well out of alignment, it is advisable to renew the frame without question or, if the amount of money available is limited, to obtain a sound replacement from a breaker's yard.

2 If the front forks have been removed from the machine, it is comparatively simple to make a quick visual check of alignment by inserting a long tube that is a good push fit in the steering head races. Viewed from the front, the tube should line up exactly with the centre line of the frame. Any deviation from the true vertical position will immediately be obvious; the steering head is a particularly good guide to the correctness of alignment when front end damage has occurred. More accurate checking must be carried out when the frame is stripped; the accompanying illustration lists the basic dimensions and angles.

11 Swinging arm rear suspension - dismantling, examination and renovation

1 Before access to the swinging arm fork is available, it is necessary to place the machine on the centre stand so that it is standing firmly on level ground. Remove both silencers (downswept exhaust system models only) and the rear wheel complete with brake drum and sprocket. Refer to Chapter 7 for the recommended procedure relating to this second task.

2 Place a container beneath the swinging arm fork pivot and remove the end cap screw from the right hand side of the pivot. Lift off the end cap, dust cover, O ring and long screw as a complete unit. Oil from the pivot will drain into the container provided.

3 Detach the right hand footrest assembly complete. Note that the earth connection is made behind one of the nuts; the zener diode need not be removed if the electrical connection is broken at the snap connector.

4 Disconnect the chain oiler pipe from its joint with the tube attached to the rear chainguard and remove the lower bolts from both rear suspension units.

5 Remove the locking screw which passes through the centre of the swinging arm pivot and threads into the pivot tube itself. Then pull out the pivot tube from the right hand side by screwing a bolt of the correct size and thread into the centre of the pivot tube and using it as a removal aid. The bolt through the centre of the front engine mounting is of the correct size and thread. The pivot tube should be a good sliding fit.

6 Lift away the swinging arm fork complete with chainguard. The left hand end cap will now be displaced, together with the left hand dust cap and O ring seal.

7 The bushes which form the bearing surface for the pivot tube are a tight fit in the inner ends of the eyes of the swinging arm fork. If play is evident between the pivot pin and the bearings, the pivot pin and bearings should be renewed as a complete set. Note that a small amount of movement at the bearing will be greatly magnified at the fork ends and will be responsible for poor handling characteristics. It may also cause the machine to fail in any machine examination test. It is important that the pivot pin is a good sliding fit, without any suspicion of play.

12 Swinging arm rear suspension - reassembly

1 Reassemble the swinging arm fork by reversing the dismantling procedure. Ensure the pivot pin and bushes are well oiled and that the pivot pin is positioned so that the tapped hole is in the 12 o'clock position to align correctly with the locking screw. The left hand end cap, dust cover and O ring are best held in position with grease so that they are not displaced before the long centre screw is engaged and tightened. Beware of over-

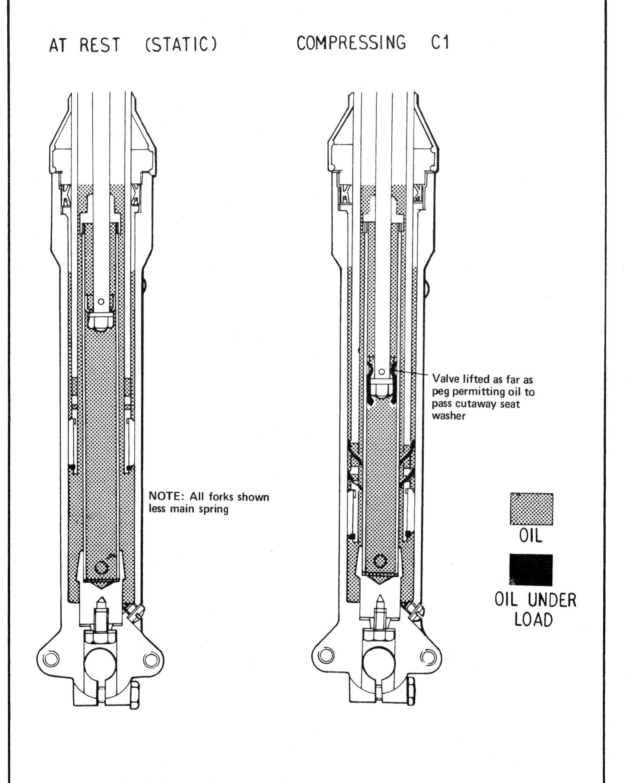

AT REST (STATIC)   COMPRESSING   C1

Valve lifted as far as
peg permitting oil to
pass cutaway seat
washer

NOTE: All forks shown
less main spring

OIL

OIL UNDER
LOAD

Fig. 6.2. Diagrammatic illustration of fork damping action (at rest and compressing)

FINAL COMPRESSION C2

EXTENDING

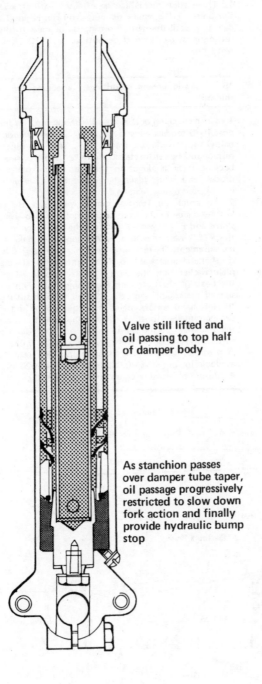

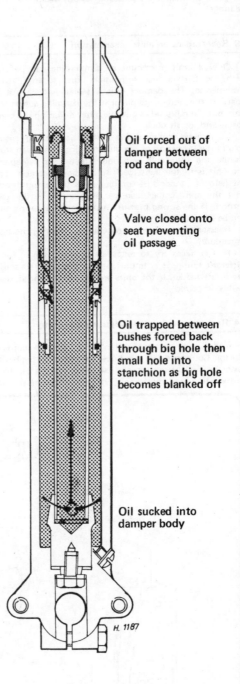

Oil forced out of damper between rod and body

Valve closed onto seat preventing oil passage

Valve still lifted and oil passing to top half of damper body

Oil trapped between bushes forced back through big hole then small hole into stanchion as big hole becomes blanked off

As stanchion passes over damper tube taper, oil passage progressively restricted to slow down fork action and finally provide hydraulic bump stop

Oil sucked into damper body

H. 1187

Diagrammatic illustration of fork damping action (Final compression and extending)

tightening the centre screw; it is very thin and will shear easily.

2 Before the grease nipple is replaced in the right hand end cap, pump the whole pivot assembly with 140 EP oil from an oil gun, until oil commences to exude from the ends of the bearings.

13 Rear suspension units - examination

1 Only a limited amount of dismantling can be undertaken with the Girling rear suspension units fitted to the Norton Commando. The damper unit is sealed and cannot be taken apart; if the unit leaks oil or if the damping action is lost, the unit must be replaced as a whole, after removing the compression spring and top shroud.

2 To remove or change the compression spring, it is necessary to grip the lower end of the unit in a vice fitted with soft clamps. Rotate the adjustable cam ring with a C or peg spanner until the unit is in the light load (solo riding) position. Then, with the help of a second person, compress the spring downwards so that the split collars can be displaced from the top of the shroud. If the spring pressure is released, the spring and shroud can be lifted away.

3 Reassemble the suspension unit by reversing the dismantling procedure.

4 If it is necessary to renew or change a spring for one of a different rating, it is important that both rear suspension units are treated alike. If the units are not well balanced, roadholding will be impaired.

14 Rear suspension units - adjusting the setting

1 The Girling rear suspension units fitted as standard to the Norton Commando have a three-position cam ring built into the lower portion of the leg, so that spring tension can be varied to suit existing load conditions. The lowest position should suit the average solo rider, under normal road conditions. When a pillion passenger is carried, the second or middle position offers a better choice and for continuous high speed work or competition events, the highest position is recommended.

2 These adjustments can be effected without need to detach the units. A C spanner or metal rod can be used to rotate the cam ring until the desired setting is obtained. It follows that the setting must be identical on both units, in the interests of good handling.

15 Isolastic engine mountings - checking and adjusting the shimming

1 When checking or adjusting the shimming, it is important that post-1970 models are supported by a strong wooden box placed below the lower main frame tubes, with the centre stand fully retracted. The centre stand cannot be used as a means of support because it bolts direct to the rear engine plates on the later models and would place the engine mountings under stress if it is subjected to any loading.

2 To check the front shimming, peel back the gaiter on the left hand side of the front engine mounting, to give access to the shims and PTFE washer. Measure the gap between the shims and the PTFE washer with a feeler gauge and make a note of the measurement. The recommended clearance is 0.010 in (0.25 mm).

3 If the clearance is too great, remove the self-locking nut and plain washer from the central mounting bolt. Align the flats on the head of the bolt with the timing casting on the right hand side of the engine and drive the bolt through sufficiently so that the left hand gaiter, engine mounting collar and PTFE washer can be removed together. Detach the dished end cap and check the thickness of the shim washers fitted. Re-shim the assembly so that the amount of free play is as close to 0.010 in (0.25 mm) as possible, using as few shims as possible. Note that shims are available in four sizes - 0.005 in, 0.010 in, 0.020 in and 0.030 in.

4 Before reassembly commences, wash and grease all the metal

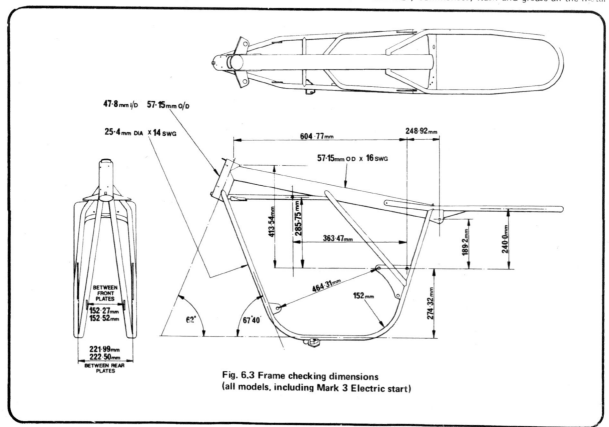

Fig. 6.3 Frame checking dimensions
(all models, including Mark 3 Electric start)

parts and check that the PTFE washer has not worn unevenly. Reassemble the component parts in the order shown in the accompanying illustration, taking care that the end cap, complete with shims, fits over the mounting body. Fit the gaiter, engine mounting collar and PTFE washer as a complete unit and tighten the self-locking nut to a torque setting of 25 ft lb.

5 A similar method is used for checking the shimming of the rear engine mounting, except that the right hand gaiter is peeled back to give access to the shims for measurement. The rear engine mounting should also have a recommended clearance of 0.010 inch. Make a note of the actual measurement if re-shimming is necessary.

6 The rear engine mounting is accessible from the right hand side for re-shimming. Remove the self-locking nut and plain washer from the central mounting bolt and drive the bolt partially through to the left until it extends some 4 inches from the left hand side. Peel the right hand gaiter off the mounting and push it downwards and rearwards complete with the engine mounting collar and PTFE washer. Lift off the dished end cap and shims, also the PTFE washer. Measure the thickness of the shims and re-adjust with the minimum number of shims so that the amount of free play on reassembly will be as close to 0.010 inch as possible. Remember to clean and grease the metal parts prior to reassembly and to check the PTFE washer for

11.5a Bolt threaded into pivot tube provides means of extraction

11.3 Chain oiler is attached to chainguard spacer

11.7 Bushes are a tight fit in swinging arm cross member

11.5 Locking bolt is long and thin; will break easily

12.2 Despite fitting of grease nipple, pivot is lubricated by oil

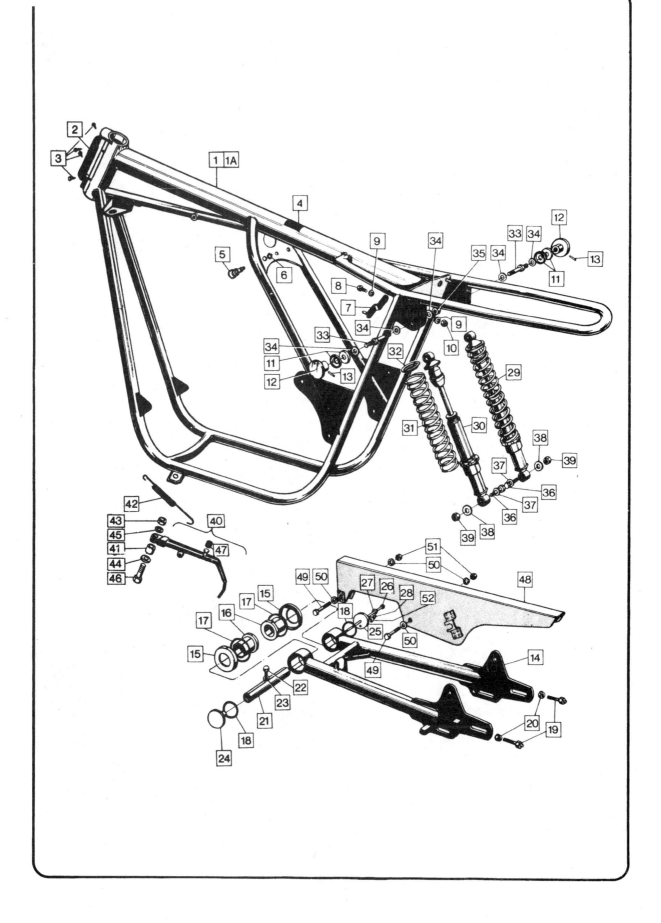

FIG. 6.4. FRAME AND REAR SUSPENSION

1 Frame - 750 cc models only
1a Frame - 850 cc models only
2 Nameplate
3 Rivet 4 off
4 Decal
5 Dzus fastner
6 Circlip for fastner
7 Panel support bracket
8 Bracket bolt
9 Bracket washer 2 off
10 Bracket nut
11 Lipped washer 4 off
12 Knurled nut for dual seat 2 off
13 Pin for nut 2 off
14 Swinging arm fork
15 Dust cover 2 off
16 Bearing bush 2 off
17 'O' ring 2 off
18 'O' ring 2 off
19 Chain adjuster 2 off
20 Chain adjuster locknut 2 off
21 Pivot spindle
22 Bolt
23 Washer
24 End cap, left hand
25 End cap, right hand
26 End cap bolt
27 Washer
28 Oil nipple
29 Suspension unit complete (Girling) 2 off
30 Damper only 2 off
31 Spring 2 off
32 Collar 2 off
33 Suspension unit top bolt 2 off
34 Suspension unit bolt washer 2 off
35 Suspension unit bottom nut 2 off
36 Suspension unit bottom bolt 2 off
37 Suspension unit bottom washer 2 off
38 Suspension unit bottom washer 2 off
39 Suspension unit bottom nut 2 off
40 Prop stand with buffer
41 Pivot spacer
42 Prop stand return spring
43 Stand nut
44 Washer
45 Spring washer
46 Pivot bolt
47 Rubber buffer
48 Chainguard
49 Chainguard bolt 2 off
50 Washer 4 off
51 Nut 2 off
52 Nipple washer

uneven wear.

7   The above-mentioned methods are recommended for checking and re-shimming the front and rear engine mountings with both mountings in situ, especially when these items receive attention during the three-monthly (2500 miles) routine maintenance inspection. When the mountings have undergone a lengthy period of service, it is preferable to remove each mounting in turn completely, a task which, in the case of the rear mounting, will necessitate dismantling the primary transmission. This will enable each complete engine mounting to be stripped and rebuilt on the workbench, affording better opportunity to clean the various component parts and renew any that are worn.

16 Isolastic engine mountings - renewing the mounting rubbers

1   Pre-1970 Commando models employed rubber mounting bushes with outer steel sleeves, which on occasion proved difficult to remove. All models manufactured during 1970 and later changed to bonded rubber bushes which are lubricated during assembly in order to obviate removal problems at a later date. No special tools are required for dismantling, but it is necessary to use Norton Villiers service tool 063971 during reassembly, to compress the rubbers prior to their insertion.

2   To renew the front engine mounting rubbers, select a metal bar which will pass into the centre of either bonded rubber bush, to the depth of the bush sleeve. If strong side pressure is applied to the bar, the bush will turn within the mounting housing and can be prised out of position. Withdraw also the central spacer tube complete with its two rubber buffers which are retained by circlips. The remaining bonded rubber bush can then be removed by side pressure with the metal bar.

3   Close examination will show whether the rubber has deteriorated, as denoted by cracks or flaking. Examine all parts for wear and replace any which are in doubtful condition.

4   To replace the first bonded rubber bush, position the tapered guide body of service tool 063971 over the end of the engine mounting and press the bush into the tapered guide after lubricating the outer edge with a rubber lubricant. As the rubber bush progresses down the guide it will be compressed to the internal diameter of the engine mounting tube and can be driven into position until the shoulder of the drift used comes against the top of the tapered guide. The first bush is now located correctly. Invert the mounting tube and slide the two buffer rubbers over the spacer tube so that they are positioned centrally with a ½ inch distance between them. They are retained in position by a circlip on each side. Insert the assembly in the mounting tube so that it rests on the first bush fitted, then fit the second bonded bush, using the service tool to compress it prior to insertion. Assembly of the front mounting rubbers is now complete.

5   A somewhat similar technique is used to dismantle and reassemble the rear engine mounting rubbers. Note that in this case two separate spacers and a centre bonded bush are employed, necessitating a slightly different approach. After the first spacer and rubber buffer assembly has been removed, it will be necessary to use the metal bar as a drift, to strike one side of the rubber portion of the bush so that the complete bonded bush will turn sideways within the mounting tube. This will make it easier to extract. When reassembling, locate the centre bonded bush first by lubricating the outer edge and pressing it into the mounting tube with a hollow tubular drift (hand pressure only) until it is located correctly 3¼ inches from each end of the tube. Position one spacer tube and rubber buffer

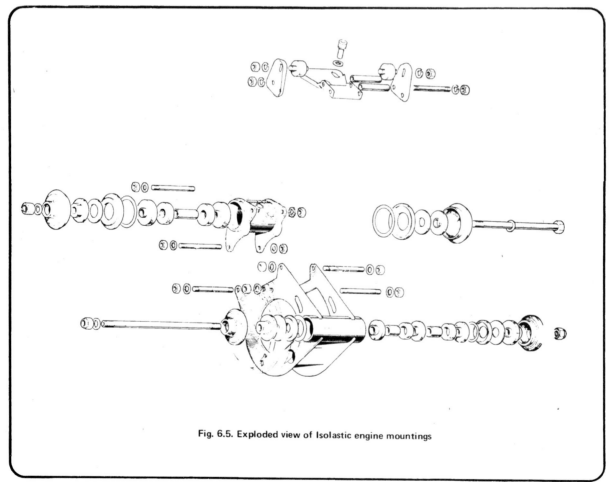

Fig. 6.5. Exploded view of Isolastic engine mountings

first, then lubricate and position the second bonded rubber bush until it abuts the spacer tube assembly on the same side of the mounting. Repeat, using an identical procedure for the replacement of the other spacer bush assembly and outer bonded bush.

17 Centre stand - examination

1 As mentioned in Section 15.1 the post-1970 models have the centre stand bolted direct to the rear engine plates. Earlier models used another arrangement in which the centre stand bolted to a cross tube, across the lower frame members. Either type of stand has a return spring to retract it, so that when the machine is pushed forward after parking, the centre stand will spring back into the fully retracted position and permit the machine to be wheeled, prior to riding.
2 The condition of the return spring and the return action should be checked frequently, also the security of the retaining nuts and bolts. If the stand drops whilst the machine is in motion, it may catch in some obstacle and unseat the rider.

18 Prop stand - examination

A prop stand which pivots from a lug welded to the lower left hand frame tube provides an additional method of parking the machine when the centre stand is not used. This too has a return spring which should have sufficient tension to cause the stand to retract immediately the weight of the machine is lifted from it. It is important that this stand is examined at regular intervals, also the nut and bolt which act as the pivot. A falling prop stand can have serious consequences if it should fall whilst the machine is on the move.

19 Footrests - examination and renovation

1 The footrests, each of which is secured to an aluminium alloy plate bolted to the main frame assembly, are malleable and will bend if the machine is dropped. Before they can be straightened, they must be detached from the alloy plate and have the rubbers removed.
2 To straighten the footrests, clamp them in a vice and apply leverage from a long tube which slips over the end. The area in which the bend has occurred should be heated to a dull cherry red with a blow lamp during the bending operation; if the footrests are bent cold, there is risk of a sudden fracture.

20 Speedometer - removal and replacement

1 A Smiths speedometer of the magnetic type is fitted to all Commando models, calibrated in miles per hour or kilometres per hour (Continental models). An internal lamp is provided for illuminating the dial and the odometer has a trip setting so that the lower of the two mileage recordings can be set to zero before a run is commenced.
2 The base of the speedometer has two studs, which permit it to be attached to a rubber mounted instrument case, bolted to the upper fork yoke. By this means it is effectively isolated from vibrations which may cause an inaccurate reading.
3 Apart from defects in the drive or the drive cable itself, a speedometer which malfunctions is difficult to repair. Fit a replacement or alternatively entrust the repair to an instrument repair specialist, bearing in mind that the speedometer must function in a satisfactory manner to meet statutory requirements.
4 If the odometer readings continue to show an increase, without the speedometer indicating the road speed, it can be assumed the drive and drive cable are working correctly and that the speedometer head itself is at fault.

21 Speedometer cable - examination and renovation

1 It is advisable to detach the speedometer drive cable from time to time in order to check whether it is adequately lubricated, and whether the outer covering is compressed or damaged at any point along its run. A jerky or sluggish speedometer movement can often be attributed to a cable fault.
2 To grease the cable, withdraw the inner cable. After removing the old grease, clean with a petrol soaked rag and examine the cable for broken strands or other damage.
3 Regrease the cable with high melting point grease and ensure that there is no grease on the last six inches, at the end where the cable enters the speedometer head. If this precaution is not observed, grease will work into the speedometer head and immobilise the movement.
4 Inspection will show whether the speedometer drive cable has broken. If so, the inner cable can be removed and replaced with another whilst leaving the outer cable in place - provided the outer cable is not damaged or compressed at any point along its run. Measure the cable length exactly when purchasing a replacement, because this measurement is critical.

22 Tachometer - removal and replacement

1 The tachometer drive is taken from the right hand end of the camshaft and emerges via a right angle take-off in front of the right hand cylinder barrel, to which the drive cable is attached.
2 It is not possible to effect a satisfactory repair to a defective tachometer head, hence replacement is necessary if the existing head malfunctions. Make sure an exact replacement is obtained; some tachometer heads work at half-speed if a different type of drive gearbox is employed.
3 The tachometer head is illuminated internally so that the dial can be read during the hours of darkness. It is mounted in an identical manner to that employed for the speedometer.

23 Tachometer drive cable - examination and renovation

Although a little shorter in length, the tachometer drive cable is identical in construction to that used for the speedometer drive. The advice given in Section 21 of this Chapter applies also to the tachometer drive cable.

24 Dualseat - removal

The seat is attached to the subframe by means of two large diameter knurled nuts which will release the dualseat when slackened. The police Interpol models have a single seat which has a similar fixing arrangement. Interstate, Roadster and Hi-Rider models have an additional metal grab rail clamped to the subframe which extends to the rear of the dualseat.

25 Steering head lock

1 A steering head lock is fitted into the top fork yoke. If the forks are turned fully to either the right or the left, they can be locked in that position as a precaution against theft.
2 Add an occasional few drops of thin machine oil to keep the lock in good working order. This should be added to the periphery of the moving drum and NOT the keyhole.

26 Cleaning - general

1 After removing all surface dirt with a rag or sponge which is washed frequently in clean water, the application of car polish or wax will restore a good finish to the cycle parts of the machine after they have dried thoroughly. The plated parts should require only a wipe with a damp rag, although it is permissible to use a chrome cleaner if the plated surfaces are badly tarnished.

2   Oil and grease, particularly when they are caked on, are best removed with a proprietary cleanser such as Gunk or Jizer. A few minutes should be allowed for the cleanser to penetrate the film of oil and grease before the parts concerned are hosed down. Take care to protect the magneto, carburettor(s) and electrical parts from the water, which may otherwise cause them to malfunction.

3   Polished aluminium alloy surfaces can be restored by the application of Solvol Autosol or some similar polishing compound, and the use of a clean duster to give the final polish.

4   If possible, the machine should be wiped over immediately after it has been used in the wet, so that it is not garaged under damp conditions which will promote rusting. Make sure to wipe the chain and if necessary re-oil it, to prevent water from entering the rollers and causing harshness with an accompanying high rate of wear. Remember there is little chance of water entering the control cables if they are lubricated regularly, as recommended in the Routine Maintenance section.

27 Sidecar attachment

The adoption of the Isolastic engine mounting system used on the Norton Commando precludes the fitting of a sidecar and for this reason no sidecar attachment lugs are provided. The manufacturer stresses that under no circumstances should a sidecar be fitted to a Norton Commando even if the attachment problems can be overcome.

28 Fault diagnosis

Symptom	Cause	Remedy
Machine is unduly sensitive to road conditions	Forks and/or rear suspension units have defective damping	Check oil level in forks. Replace rear suspension units.
Machine tends to roll at low speeds	Steering head bearings overtight or damaged	Slacken bearing adjustment. If no improvement, dismantle and inspect bearings.
Machine tends to wander, steering is imprecise	Worn swinging arm bearings or excess clearances in engine mountings	Check and if necessary renew bearings and/or re-shim engine mountings.
Fork action stiff	Fork legs have twisted in yokes or have been drawn together at lower ends	Slacken off spindle nut clamps, pinch bolts in fork yokes and fork top nuts. Pump forks several times before retightening from bottom. Is distance piece missing from fork spindle?
Forks judder when front brake is applied	Worn fork bushes Steering head bearings too slack	Strip forks and replace bushes. Re-adjust, to take up play.
Wheels out of alignment	Frame distorted as result of accident damage	Check frame alignment after stripping out. If bent, specialist repair is necessary.

Chapter 7 Wheels, brakes and tyres

Contents

Specifications

Wheel sizes
   Front and rear ... ...   19 inch diameter, all models. WM2 - 19 rim

Tyre sizes
   Front and rear ... ...   4.10 inch x 19 inch, all models

Tyre pressures
   Front ... ...   26 psi, all models Avon GP
                  22 psi, all models Dunlop K81, TT100
   Rear ... ...   26 psi, all models Avon GP
                  24 psi, all models Dunlop K81, TT100

Brakes
   Front ... ...   8 inch internal diameter, twin leading shoe
   Rear ... ...   7 inch diameter, single leading shoe
   Brake lining area ... ...   18.69 sq in (474.776 sq mm)
   Front disc brake: disc diameter ... ...   10.70 in (271.7 mm)
   Pad thickness (including backing) ... ...   0.38 - 0.37 in (9.652 - 9.398 mm)
   Brake fluid ... ...   Lockheed series 329 Hydraulic Fluid for disc brakes (Complies with US Safety Standard 116)

Wheel bearings
   Type ... ...   Ball journal
   Size ... ...   17 mm x 40 mm x 12 mm (post-1971)

Brake drum bearing
   Size ... ...   NM 17721 (fitted to post-1971 models only)

1 General description

All Norton Commando models are fitted with 19 inch diameter wheels, carrying tyres of 4.10 inch section. The rear wheel is of the quickly-detachable type and can be removed from the frame without need to disturb the final drive sprocket and brake drum.

The Hi-Rider, Fastback and some Interpol models are fitted with drum brakes of the internally-expanding type, front and rear. The front drum is 8 inches in diameter, the rear drum 7 inches. The front brake employs a twin leading shoe arrangement. Other models, including those of 850 cc capacity, have a Lockheed front disc brake which is hydraulically operated, to provide braking efficiency which is in keeping with engine performance.

2 Front wheel - examination and renovation

1 Place the machine on the centre stand so that the front wheel is raised clear of the ground. Spin the wheel and check for rim alignment. Small irregularities can be corrected by tightening the spokes in the affected area, although a certain amount of experience is necessary if over-correction is to be avoided. Any 'flats' in the wheel rim should be evident at the same time. These are more difficult to remove with any success and in most cases the wheel will have to be rebuilt on a new rim. Apart from the effect on stability, especially at high speeds, there is much greater risk of damage to the tyre beads and walls if the machine is ridden with a deformed wheel.
2 Check for loose or broken spokes. Tapping the spokes is the best guide to tension. A loose spoke will produce a quite different note that should be tightened by turning the nipple in an anticlockwise direction. Always recheck for run-out by spinning the wheel again.

3 Front drum brake assembly - examination, renovation and reassembly

1 The front wheel can be removed from the forks and the brake assembly complete with brake plate detached by following the procedure given in Chapter 6, Section 2.4, when the machine is on the centre stand.
2 Before dismantling the brake assembly, examine the condition of the brake linings. If they are wearing thin or unevenly, they must be replaced. The angle of the brake operating arms will give the best indication of the degree of wear, if checked when the brake is applied.
3 To remove the brake shoes from the brake plate, first detach the support plate or separate metal strips which bolt to the brake shoe pivot or torque stop pins. They are secured with a tab washer. Then position the brake operating cams so that the shoes are in the fully-expanded position and pull the shoes apart whilst lifting them upwards and bringing them together in the form of a V. When they are clear of the brake plate, the return springs should be detached and the shoes separated.
4 It is possible to replace the brake linings fitted with rivets and not bonded on, as is the current practice. Much will depend on the availability of the original type of linings; service-exchange brake shoes with bonded-on linings may be the only practical form of replacement.
5 If new linings are fitted by rivetting, it is important that the rivet heads are countersunk, otherwise they will rub on the brake drum and be dangerous. Make sure the lining is in very close contact with the brake shoe during the rivetting operation; a small C clamp of the type used by carpenters can often be used to good effect until all the rivets are in position. Finish off by chamfering off the end of each shoe, otherwise fierce brake grab may occur due to the pick-up of the leading edge of each lining.
6 Before replacing the brake shoes, check that the brake operating cams are working smoothly and not binding in their pivots. The cams can be removed for greasing by unscrewing the

nut on the end of each operating arm, after making the arms and the brake plate so that they are replaced in an identical position. Draw the operating arm off the squared end of each operating cam and do not slacken or remove the rod which joins them, or the relationship between the two arms will be lost and will have to be reset on assembly. The operating cams will withdraw from the inside of the brake plate, permitting the shaft and the bush in which each cam operates to be cleaned and greased.
7 Check the inner surface of the brake drum, on which the brake shoes bear. The surface should be smooth and free from indentations, or reduced braking efficiency is inevitable. Remove all traces of brake lining dust and wipe the surface with a rag soaked in petrol to remove any traces of grease or oil.
8 To reassemble the brake shoes on the brake plate, fit the return springs and force the shoes apart, holding them in a V formation. If they are now located with the brake operating arm and pivot, they can usually be snapped into position by pressing downward. Do not use excessive force, or the shoes may be distorted permanently. Do not omit to replace the support plates or strips.

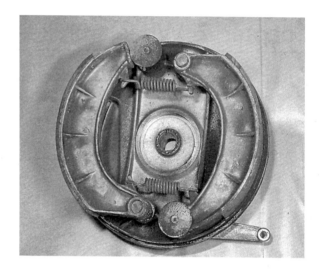

3.1 Front brake assembly is of the twin leading shoe type

3.8 When replacing the wheel in the forks, the brake stop must align as shown

4 Front disc brake assembly - examination, renovation and reassembly

1 The brake disc attached to the front wheel rarely requires any attention. Renewal is necessary only if the surface is scored or damaged. Rust build-up may occur on the cast iron discs fitted to early models, which may prevent entry of the disc when new friction pads are fitted. Provided the rust is not too deep, it can be removed with fine, smooth emery paper. Later machines employ a stainless steel disc, which obviates this problem completely.

2 The disc is bolted to the wheel hub with five bolts and tab washers. If the front wheel is removed from the forks, as described in Chapter 6, Section 2.3, the disc will be freed when the bolts are withdrawn. In this instance the machine can be supported on the centre stand. There is no necessity to detach the hydraulic brake system or for that matter to dismantle any part of it. The disc will pull clear of the friction pads in the caliper. It is, however, advisable to place a clean spacer, such as a piece of wood or metal, between the pads to prevent them being ejected if the front brake is unintentionally applied. This precaution is not necessary if the caliper piston assemblies are to receive attention.

3 The friction pads will lift out of the brake caliper if they are turned slightly. Inspect the friction faces for excessive wear, uneven wear or scoring. Renew both pads if there is any doubt about the condition of either one. Pads should always be renewed in pairs, never singly.

4 Clean out the recesses into which the pads fit and the exposed end of the pistons, using a soft brush. Do not on any account use a solvent cleanser or a wire brush. Finish off by giving the piston faces and the friction pad recesses just a smear of brake fluid.

5 If it is found that the pistons do not move freely or are seized in position, the caliper is in urgent need of attention and must be removed, drained and overhauled. Seek advice from a professional experienced with motor cycle disc brakes. If a piston is seized, the only satisfactory course of action is renewal of the complete brake caliper unit.

6 To remove the brake caliper unit from the machine, unscrew the union where the hydraulic fluid pipe enters the unit and drain the complete hydraulic system into a clean container. Never re-use brake fluid. The caliper unit can now be detached from its mounting on the lower left hand fork leg by removing the two retaining bolts.

7 If the caliper unit shows evidence of brake fluid leakage, accompanied by the need to top up the hydraulic fluid reservoir at regular intervals, the piston seals require renewal. This is a simple task which is carried out as follows: Remove the front wheel complete with disc as described in Chapter 6, Section 2.3, after positioning the machine on the centre stand. There is no necessity to detach the hydraulic brake system. Lift both friction pads out of position and mark the friction faces so that they are replaced in an identical position. Drain the hydraulic system by placing a clean receptacle below the unit to catch the hydraulic fluid and squeezing the handlebar lever so that both pistons are expelled to release the fluid. Unscrew the caliper end plug which has two peg holes and will require the use of the correct peg spanner tool because it is a tight fit. Remove the two pressure seals from their respective grooves, using a blunt nosed tool to ensure the grooves are not damaged in any way.

8 Wet the new seals with hydraulic fluid and insert the first seal into the innermost bore, making sure it has seated correctly. The diameter of the seal is larger than that of the groove into which it fits, to that a good interference fit is achieved. Furthermore, the sections of the seal and seal groove are different to ensure the sealing edge is proud of the groove. Wet one of the pistons with hydraulic fluid and insert it through the outer cylinder (uncovered by removal of the caliper end plug) so that it passes through into the innermost cylinder bore, closed end facing inward. Check that it enters the seal squarely and leave it protruding approximately 5/16 inch (8 mm) from the mouth of the inner bore.

9 Fit the seal and piston in the outer bore of the caliper unit using an identical procedure. Fit a new O ring seal and replace the end plug, tightening it to a torque setting of 26 lb ft. Replace the friction pads in their original positions after checking that all traces of fluid used to lubricate the various components during assembly have been removed, replace the front wheel and refill the master cylinder reservoir with the correct grade of hydraulic fluid. It will be necessary to bleed the system before the correct brake action is restored by following the procedure described fully in Section 6.

10 Note that all these operations must be carried out under conditions of extreme cleanliness. The brake caliper unit must be cleaned thoroughly before dismantling takes place. If particles of grit or other foreign matter find their way into the hydraulic system there is every chance that they will score the precision made parts and render them inoperative, necessitating expensive replacements.

5 Master cylinder - examination and replacing seals

1 The master cylinder and hydraulic fluid reservoir take the form of a combined unit mounted on the right hand side of the handlebars, to which the front brake lever is attached. The master cylinder is actuated by the front brake lever and applies hydraulic pressure through the system to operate the front brake when the handlebar lever is manipulated. The master cylinder pressurises the hydraulic fluid in the pipe line which, being incompressible, causes the pistons to move within the brake caliper unit and apply the friction pads to the brake disc. It follows that if the piston seals of the master unit leak, hydraulic fluid will be lost and the braking action rendered much less effective.

2 Before the master unit can be removed and dismantled, the system must be drained. Place a clean container below the brake caliper unit and attach a plastic tube from the bleed screw of the caliper unit to the container. Open the bleed screw one complete turn and drain the system by operating the brake lever until the master cylinder reservoir is empty. Close the bleed screw and remove the pipe.

3 To gain access to the master cylinder, disconnect the front brake stop lamp switch by pulling off the spade terminal connections. Lift away the switch cover and detach the hose from the master cylinder by unscrewing the union joint. Remove the four screws securing the unit to the handlebars by means of a split clamp and withdraw the unit complete with integral reservoir.

4 Remove the reservoir cap and bellows seal from the top of the reservoir. Remove the front brake stop lamp switch by unscrewing it from the master cylinder body. Release the handlebar lever by withdrawing the pivot bolt. The rubber boot over the master cylinder piston is retained in position by a special circlip having ten projecting ears. If three or four adjacent ears are lifted progressively, the circlip can gradually be lifted away until it clears the mouth of the bore and is released completely. It will most probably come away with the piston together with the secondary cup.

5 Remove the primary cup washer, cup spreader and bleed valve assembly which will remain within the cylinder bore. They are best displaced by applying gentle air pressure to the hose union bore.

6 Examine the cylinder bore for wear in the form of score marks or surface blemishes. If there is any doubt about the condition of the bore, the master cylinder must be renewed. Check the brake operating lever for pivot bore wear, cracks or fractures, the hose union and switch threads, and the piston for signs of scuffing or wear. Finally, check the brake hose for cuts, cracks or other signs of deterioration.

7 Before replacing the component parts of the master cylinder, wash them all in clean hydraulic fluid and place them in order of assembly on a clean, dust-free surface. Do not wipe them with a fluffy rag; they should be allowed to drain. Particular attention should be given to the replacement primary and secondary cup

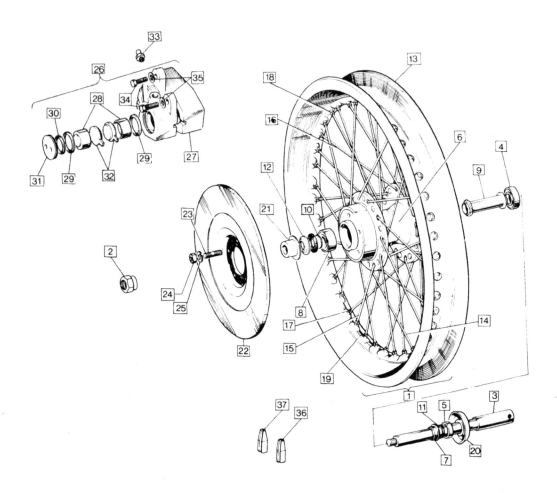

FIG. 7.1. FRONT WHEEL, DISC BRAKE TYPE

1   Rim and hub assembly
2   Wheel spindle nut
3   Wheel spindle
4   Left hand wheel bearing
5   Bearing lockring
6   Hub shell
7   Spacer
8   Right hand wheel bearing
9   Spacer
10 Washer
11 Seal 2 off
12 Washer
13 Rim - size WM2-19
14 Spoke, right hand inner 10 off
15 Spoke, right hand outer 10 off
16 Spoke, left hand inner 10 off
17 Spoke, left hand outer 10 off
18 Spoke nipple, right hand 20 off
19 Spoke nipple, left hand 20 off

20 Dust cover
21 Right hand spacer
22 Disc
23 Stud 5 off
24 Nut 5 off
25 Washer 5 off
26 Caliper assembly
27 Caliper (not supplied separately)
28 Piston 2 off
29 Seal 2 off
30 Seal 2 off
31 End plug
32 Friction pads 2 off
33 Bleed nipple
34 Screw 2 off
35 Spring washer 2 off
36 Balance weight, 17 grammes
37 Balance weight, 25 grammes

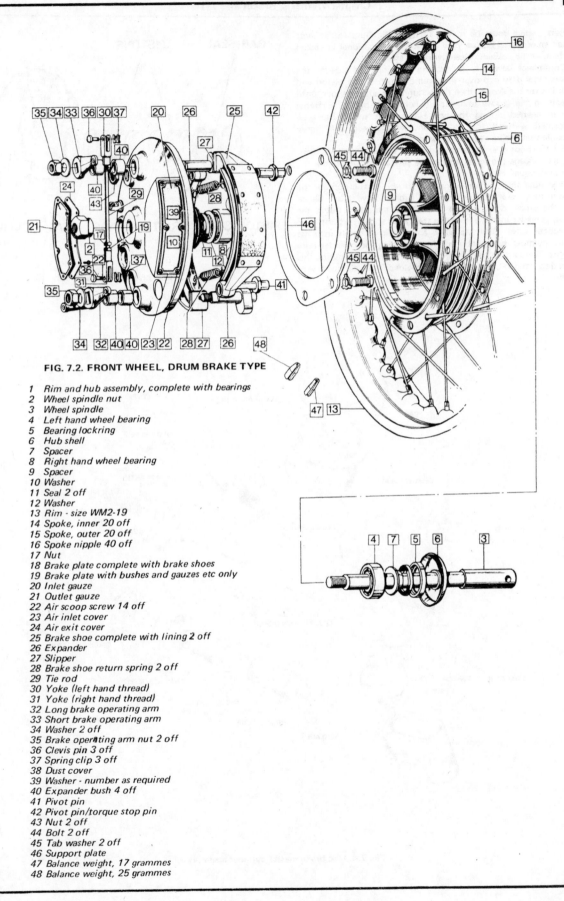

FIG. 7.2. FRONT WHEEL, DRUM BRAKE TYPE

1 Rim and hub assembly, complete with bearings
2 Wheel spindle nut
3 Wheel spindle
4 Left hand wheel bearing
5 Bearing lockring
6 Hub shell
7 Spacer
8 Right hand wheel bearing
9 Spacer
10 Washer
11 Seal 2 off
12 Washer
13 Rim - size WM2-19
14 Spoke, inner 20 off
15 Spoke, outer 20 off
16 Spoke nipple 40 off
17 Nut
18 Brake plate complete with brake shoes
19 Brake plate with bushes and gauzes etc only
20 Inlet gauze
21 Outlet gauze
22 Air scoop screw 14 off
23 Air inlet cover
24 Air exit cover
25 Brake shoe complete with lining 2 off
26 Expander
27 Slipper
28 Brake shoe return spring 2 off
29 Tie rod
30 Yoke (left hand thread)
31 Yoke (right hand thread)
32 Long brake operating arm
33 Short brake operating arm
34 Washer 2 off
35 Brake operating arm nut 2 off
36 Clevis pin 3 off
37 Spring clip 3 off
38 Dust cover
39 Washer - number as required
40 Expander bush 4 off
41 Pivot pin
42 Pivot pin/torque stop pin
43 Nut 2 off
44 Bolt 2 off
45 Tab washer 2 off
46 Support plate
47 Balance weight, 17 grammes
48 Balance weight, 25 grammes

washers, which must be soaked in hydraulic fluid for at least fifteen minutes to ensure they are supple. Occasional kneading will help in this respect. Clean hands are essential.

8   Commence assembly by placing the unlipped side of the hollow secondary cup over the ground crown of the piston and work it over the crown, then the piston body and shoulder, until it seats in the groove immediately below the piston. Extreme care is needed during the entire operation which must be performed by hand. If a tool of any kind is used, it is highly probable that the lip seal will be damaged.

9   Fit the boot over the piston, open end toward the secondary cup, and engage the upper end in the piston groove, so that the boot seats squarely.

10 Assemble the trap valve spring over the plastic bobbin; the bobbin must seat squarely in the rubber valve base. Check that the small bleed hole is not obstructed, then insert the plastic spreader in the end of the trap valve spring furthest from the rubber valve. Offer the valve and spring assembly into the master cylinder bore, valve end first, keeping the cylinder bore upright. The casting can be held in a soft jawed vice during this operation to facilitate assembly, provided it is clamped very lightly.

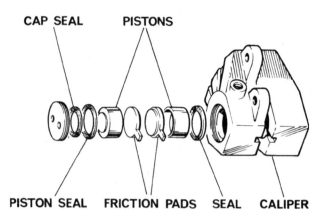

Fig. 7.3. Disc brake caliper assembly

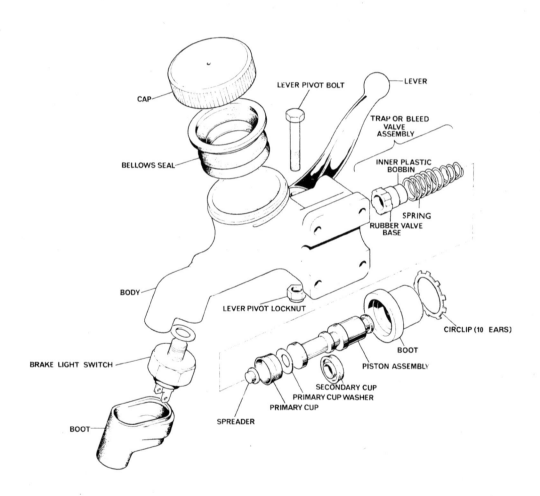

Fig. 7.4. Disc brake master cylinder assembly

11 Insert the primary cup into the bore with the belled-out end innermost. Insert the primary cup washer with the dished portion facing the open end of the bore. Insert the piston, crown end first, into the bore and locate the ten ear circlip over the boot, so that the set of the ears faces away from the cylinder. Apply pressure with a rotary motion and check that the lip of the secondary cup enters the cylinder bore without damaging the lip. Maintain pressure and engage the lower shoulder of the boot with the counterbore of the cylinder which acts as its seating. Work the boot retaining circlip into position, whilst still maintaining pressure on the piston assembly.

12 Still maintaining pressure on the piston assembly, feed the brake lever into position at the fulcrum slot and align the pivot holes so that the pivot bolt can be replaced and locked with the locknut.

13 It must be emphasised that reassembly of the master cylinder is a very tricky and delicate operation in which great care has to be exercised to ensure the replacement seals are not damaged. The assistance of another person during reassembly is advisable since without previous experience, it is difficult to maintain the required pressure on the piston assembly whilst the final operations are completed.

14 Replace the master cylinder on the handlebars and refill the reservoir with hydraulic brake fluid. The system must now be bled.

6 Front disc brake - bleeding the hydraulic system

1 As mentioned earlier, brake action is impaired or even rendered inoperative if air is introduced into the hydraulic system. This can occur if the seals leak, the reservoir is allowed to run dry or if the system is drained prior to the dismantling of any component part of the system. Even when the system is refilled with hydraulic fluid air pockets will remain and because air will compress, the hydraulic action is lost.

2 Check the fluid content of the reservoir and fill almost to the top. Remember that hydraulic brake fluid is an excellent paint stripper, so beware of spillage, especially near the petrol tank.

3 Place a clean glass jar below the brake caliper unit and attach a clear plastic tube from the caliper bleed screw to the container. Place some clean hydraulic fluid in the container so that the pipe is always immersed below the surface of the fluid.

4 Unscrew the bleed screw one complete turn and pump the handlebar lever slowly. As the fluid is ejected from the bleed screw the level in the reservoir will fall. Take care that the level does not drop too low whilst the operation continues, otherwise air will re-enter the system, necessitating a fresh start.

5 Continue the pumping action with the lever until no further air bubbles emerge from the end of the plastic pipe. Hold the brake lever against the handlebars and tighten the caliper bleed screw. Remove the plastic tube AFTER the bleed screw is closed.

6 Check the brake action for sponginess, which usually denotes there is still air in the system. If the action is spongy, continue the bleeding operation in the same manner, until all traces of air are removed.

7 Bring the reservoir up to the correct level of fluid (within ½ inch of the top of the reservoir) and replace the bellows seal and cap. Check the entire system for leaks. Recheck the brake action.

8 Note that fluid from the container placed below the brake caliper unit whilst the system is bled, should not be re-used.

7 Front wheel bearings - examination and renovation

Drum brake wheel only

1 When the brake plate and brake assembly has been withdrawn, unscrew the locking ring which retains the left hand bearing. The locking ring has a right hand thread and can be unscrewed by using Norton Villiers service tool 063965 - a peg spanner - or by the use of a pin punch and hammer. If the ring proves difficult to remove, warm the hub and apply the peg spanner or punch in different holes.

2 After the locking ring has been unscrewed, lift out the felt seal and distance piece. Then insert the front wheel spindle from the right hand (brake drum) side of the hub to drive the right hand double row bearing inwards whilst at the same time displacing the left hand single row bearing from the hub. A rawhide mallet should be used to drive the spindle so that the displaced bearing only just clears the hub. If this precaution is not observed, there is risk of damaging the spacer.

3 Working from the other side of the wheel, enter the wheel spindle through the spacer tube which is still within the hub and drive the double row bearing outwards again, whilst holding the whole assembly in a central position. The bearing will be displaced from the hub, together with the felt retaining washer, felt seal, outer washer and spacer.

4 Remove all the old grease from the hub and bearings and give the latter a final wash in petrol. Check the bearings for play or signs of roughness when they are turned. If there is any doubt about their condition, play safe and renew them. A new bearing has no discernible play.

5 Before replacing the bearings, first pack the hub with new, high melting point grease and the bearings themselves. Locate the single row bearing first in the left hand side of the hub, then fit the distance washer (plain side to bearing) the felt seal and the locking ring. Tighten the ring fully with either the peg spanner or the pin punch.

6 Insert the bearing spacer tube from the right, small end first, so that it presses against the bearing already positioned. Pass the front wheel spindle through the double row bearing and enter this bearing into the right hand end of the hub. Push the wheel spindle further into the hub so that it passes through the spacer tube and through the centre of the left hand bearing, until the head abuts the end of the double row bearing. Drive the end of the spindle so that the double row bearing enters the hub fully and abuts the spacer tube. Refit the smaller felt retaining washer, the felt seal and the large steel washer. This latter washer is either peened or pushed into position, depending on the type fitted.

Disc brake wheel only

1 A somewhat different procedure is required for the disc brake wheel, which has an aluminium alloy hub. Commence by unscrewing the locking ring on the left hand side of the hub, using either Norton Villiers service tool 063965 or a pin punch and hammer. If the ring proves difficult to remove, warm the hub and apply the peg spanner or pin punch in different holes.

2 After the locking ring has been unscrewed, lift away the seal and spacer. Then heat the hub with hot water to a temperature not exceeding 100°C. Insert a bar into the bearing spacer tube and force the tube out of position sufficiently to permit a drift to make contact with the inner race of the bearing to be displaced. Drive the single row bearing outwards, whilst gradually working the spacer tube across to the other side of the hub so that the drift will make better contact with the outgoing bearing. Continue until the single row bearing is displaced from the hub.

3 Take out the spacer tube which should be replaced if damaged as a result of the preceding operation. Then using a suitable size of drift, drive the double row bearing outwards from the inside of the hub, taking with it the washers and felt seal. Refer to the previous instructions for the examination and replacement of the drum brake wheel bearings (paragraph 4 onwards) which apply equally well to the disc brake wheel.

8 Rear wheel - examination, removal and renovation

1 Before removing the rear wheel, check for rim alignment, damage to the rim and loose or broken spokes by following the procedure described in Section 2 of this Chapter.

2 To remove the rear wheel without disturbing the final drive chain, place the machine on the centre stand so that the rear wheel is clear of the ground. Disconnect the speedometer drive

cable from the gearbox through which the rear wheel spindle passes, then unscrew the spindle from the right hand side of the machine (right hand thread). When the spindle is withdrawn, the spacer will fall free and the gearbox can be pulled off the end of the rear wheel hub.

3 To separate the wheel from the brake drum and sprocket, insert a lever between the wheel hub and the back of the brake drum to disengage the three projecting 'paddles' from the cush drive buffers in the hub.

4 Pre-1971 models have a slightly different arrangement. Before the wheel spindle is slackened and withdrawn, it is necessary to remove the three rubber blanking plugs from the right hand side of the hub and the three sleeve nuts contained within the tunnels blanked off by the plugs. The wheel can then be pulled off the studs which project from the rear of the brake drum, after the rear wheel spindle is slackened and withdrawn from the right.

5 Occasions arise when it is necessary to remove the rear wheel complete with brake drum and sprocket, in which case a different procedure is necessary. Commence by supporting the machine on the centre stand, so that the rear wheel is clear of the ground. Then disconnect the rear chain at the split link, a task made easier if the link is positioned in the teeth of the rear wheel sprocket.

6 Disconnect the rear brake cable rod by removing the adjuster nut and pulling the rod through the trunnion in the end of the brake operating arm. Disconnect the speedometer drive cable from the gearbox through which the rear wheel spindle passes, then slacken and withdraw the wheel spindle from the right hand side of the machine. The spacer will fall free when the spindle is withdrawn and the speedometer gearbox can be pulled off the end of the hub.

7 Unscrew and remove the left hand spindle nut and pull the wheel over to the right hand side of the swinging arm fork so that the torque stop of the brake plate will disengage from the slotted lug of the left hand fork tube. The wheel is now free to be lifted from the rear of the machine.

9 Rear wheel - dismantling, examining and reassembling the hub

1 Before the hub can be dismantled further, it is first necessary to separate the brake assembly from the brake drum and then the brake drum from the wheel hub. The cush drive arrangement fitted to all post-1970 models permits the brake drum to be lifted away from the hub without further dismantling, by disengaging the 'paddles' which project from the rear of the brake drum.

2 In the case of pre-1971 models it is necessary to remove the three blanking plugs from the right hand end of the hub and unscrew the three sleeve nuts contained within the tunnels blanked off by the plugs. The studs which project from the back of the brake drum are then freed, permitting the brake drum to be lifted away.

3 To dismantle the rear hub, unscrew the bearing locking ring found on the right hand (speedometer drive) end of the hub. The ring has a LEFT HAND thread, and it is necessary to use either Norton Villiers service tool 063965 or a pin punch and hammer. When the ring has unscrewed, lift out the distance piece and felt seal.

4 Replace the thick washer on the rear wheel spindle and the right hand spacer which fits between the speedometer gearbox and the fork end. Insert the spindle through the bearing in the left hand (brake drum) end of the hub, with the washer and spacer against the spindle head, then drive the bearing into the hub as far as it will go by hitting the end of the spindle with a rawhide mallet. This will commence the displacement of the right hand bearing from the hub. Withdraw the rear wheel spindle and its attachments, then insert the front wheel spindle in its place, threaded end foremost, from the same side. If the spindle is held horizontally and the end tapped with the mallet, the right hand bearing and bearing spacer will be displaced completely from the hub.

5 Insert the rear wheel spindle and thick washer into the wheel

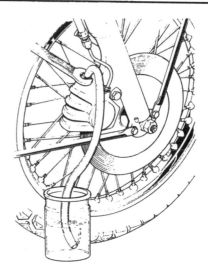

Fig. 7.5. Bleeding hydraulic disc brake

8.2 Rear wheel spindle unscrews from right, then pulls out

8.2a Speedometer gearbox pulls off wheel hub

8.7 Left hand spindle nut must be removed if wheel is to be detached with sprocket and brake drum attached

hub from the opposite (right hand) side and drive the left hand bearing outwards from the hub, complete with the felt retaining washer, felt seal and dished washer.

6 Some minor variations will occur with regard to the type of bearing fitted to the left hand side of the hub, depending on the year of manufacture of the machine. Early models have a double row bearing in this location, whereas later models have only a single row bearing. This is because the later models carry a cush drive assembly in the hub, necessitating the addition of an extra double row bearing within the brake drum itself to ensure rigidity of the modified assembly.

7 Remove all old grease from the hub and bearings, then check the bearings for signs of play or roughness as described in Section 7.4.

8 To reassemble the hub, fit the right hand bearing first so that the felt seal and locking ring can be fitted and the latter tightened fully. Special care is necessary to ensure the slots which transmit the drive to the speedometer gearbox are not damaged. Then insert the spacer tube into the hub from the left, noting that pre-1971 models employ a spacer with ends of different lengths so that the longer end will locate with the bearing on the right hand side of the hub. Insert the left hand bearing and drive it into the hub, applying load only to the outer race. Fit the felt retaining washer, felt seal and dished washer, which is either peened or pushed into position, depending on the type fitted.

10 Brake drum bearing and cush drive assembly - examination and renovation

1971-on models only

1 Lay the brake drum face downwards, after detaching the brake plate and brake assembly. Remove the spacer from the centre of the bearing within the brake drum; it is a tight fit and may have to be held securely whilst the brake drum is driven away.

2 Prise out the lipped washer which covers the felt seal, then remove the felt seal and the felt retaining washer. This will expose the circlip which retains the bearing, which should be removed with a pair of circlip pliers.

3 Turn the brake drum the other way up and rest it across two wooden blocks so that the dummy spindle can be driven inwards into the drum with a rawhide mallet to displace the inner felt seal and then the double row bearing.

4 Wash the bearing in petrol to remove all old grease and oil, then check it for roughness or play. If there is any doubt about its condition, play safe and renew. Repack the bearing with grease, even if a new bearing is to be fitted.

5 Check, examine and replace, if necessary, the bearing seals and retainers, especially if any were damaged during the dismantling operation. Make sure the shock absorber paddles are

secure in the brake drum, since they transmit the drive from the engine.

6 Fit the inner washers in the brake drum first, then the dummy spindle. Complete reassembly by reversing the dismantling procedure.

7 Before refitting the brake drum to the wheel, check that the PTFE buffers on which the paddles bear are in good order and do not require replacement.

8 Note that the rear wheel sprocket is an integral part of the brake drum. If the sprocket teeth are worn, chipped, broken or hooked, the brake drum must be renewed, preferably in conjunction with a new chain and gearbox final drive sprocket.

11 Rear brake assembly - examination, renovation and reassembly

1 The rear brake assembly, which is of the drum type in all cases, can be withdrawn complete with brake plate, after the rear wheel has been withdrawn from the frame. It is similar in construction to the front wheel drum brake, apart from the fact that it is of the single leading shoe type. In consequence, there is only one operating cam and arm.

2 Follow an identical procedure for dismantling, examining and reassembling the brake to that described in Section 3 of this Chapter.

11.1 Rear brake is of single leading shoe type

12 Adjusting the front brake

1 All models fitted with a drum front brake employ a twin leading shoe arrangement which gives maximum braking effieiency. The cable adjusters are provided, one at the end of the handlebar lever and the other on the brake plate. Adjustment is a matter of individual requirements, but it should not be possible for the end of the brake lever to touch the handlebars before the brake is applied fully, or even approach very close. On the other hand, the brake shoes should not rub on the brake drums when the brake is not in use. Apart from causing reduced performance, the brake shoes may overheat and give rise to brake fade, rendering the brake useless in an emergency. Turn either adjuster outwards (anticlockwise) to take up wear which would otherwise cause increasing slackness in the brake cable.

2 The screwed operating rod which joins the two brake operating arms of the front brake should not require attention unless the setting has been disturbed. It is imperative that the leading edge of each brake shoe comes into contact with the brake drum simultaneously, if maximum braking efficiency is to be achieved. Check by detaching the clevis pin from the eye of one end of the threaded rod, so that the brake operating arms can be applied independently. Operate each arm at a time and note when the brake shoe first commences to touch the brake drum, with the wheel spinning. Make a mark to show the exact position

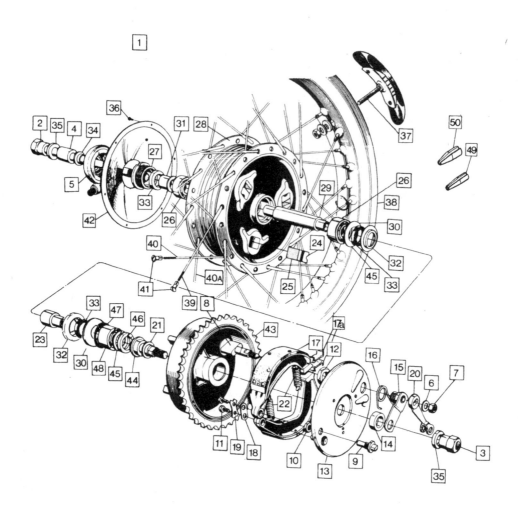

FIG. 7.6. REAR WHEEL

1	Rim and hub assembly
2	Wheel spindle
3	Wheel spindle nut
4	Spacer
5	Speedometer drive gearbox
6	Washer
7	Nut
8	Brake cam
9	Torque stop pin
10	Torque stop pin nut
11	Bolt 2 off
12	Cam bearing nut
13	Brake plate assembly
14	Brake plate outer spacer
15	Cam bearing and stay
16	Operating arm return spring
17	Brake shoe complete with lining 2 off
17A	Brake shoe slipper 2 off
18	Plate
19	Washer
20	Brake operating arm
21	Washer
22	Brake shoe return spring 2 off
23	Left-hand bearing spacer
24	Cush drive buffer - thick 3 off
25	Cush drive buffer - thin 3 off
26	Wheel bearing 2 off
27	Lockring
28	Hub shell
29	Inner bearing spacer
30	Felt retaining washer 2 off
31	Spacer
32	Dished washer 2 off
33	Seal 3 off
34	Spacer, speedometer drive gearbox
35	Spindle washer 2 off
36	Disc screw 6 off
37	Security bolt
38	Pim - size WM2-19
39	Spoke 20 off
40	Spoke, inner left-hand 10 off
40A	Spoke, inner right-hand 10 off
41	Nipple 40 off
42	Rear hub disc
43	Brake drum and sprocket
44	Dummy spindle
45	Felt retaining washer
46	Felt seal
47	Circlip
48	Brake drum bearing
49	Balance weight, 17 grammes
50	Balance weight, 25 grammes

of each operating arm when this initial contact is made. Replace the clevis pin and check that the marks correspond when the brake is applied in similar fashion. If they do not, withdraw the clevis pin and use the adjuster to either increase or decrease the length of the rod (top thread is left hand) until the marks correspond exactly. Replace the clevis pin and do not omit the split pin through the end which retains it in position. Re-check the brake lever adjustment before the machine is tried on the road.

3   Check that the brake pulls off correctly when the handlebar lever is released. Sluggish action is usually due to a poorly lubricated brake cable, broken return springs or a tendency for the brake operating cams to bind in their bushes. Dragging brakes affect engine performance and can cause severe over-heating of both the brake shoes and wheel bearings.

4   No brake adjustment is necessary in the case of models fitted with a disc front brake. The hydraulic system is self-compensating. A non-adjustable stop lamp switch is built into the master cylinder unit.

13 Adjusting the rear brake

1   The rear brake is of the single leading shoe type and has a single adjuster at the end of the brake operating arm. Turn the adjuster nut inwards (clockwise) to decrease the brake pedal movement and take up excess slack in the cable. The amount of free movement in the brake pedal is a matter of personal preference, but it should not be necessary to depress the pedal far before the brake is applied fully. Beware of adjusting the brake too closely so that the linings are still in rubbing contact with the brake drum.

2   After adjusting the rear brake, it is sometimes necessary to adjust the stop lamp switch, so that the stop lamp lights up at the correct time. The switch is bolted to a lug on the brake pedal and has a mounting plate with elongated slots, so that it can be either raised or lowered to actuate the stop lamp either earlier or later. The recommended switch depression is 1/32 inch (0.08 mm).

14 Final drive chain - examination, lubrication and adjustment

1   Except on a few models, the final drive chain does not have the benefit of full enclosure or positive lubrication which is afforded to the primary drive chain. In consequence, it will require attention from time to time, particularly when the machine is used on wet or dirty roads.

2   Chain adjustment is correct when there is approximately ¾ inch play in the middle of the run. Always check at the tightest spot on the chain run, under load.

3   If the chain is too slack, adjustment is effected by slackening the rear wheel spindle and nut, then pushing the wheel backwards by means of the chain adjusters at the end of the rear fork. Make sure each adjuster is turned an equal amount, so that the rear wheel is kept centrally-disposed within the frame. When the correct adjusting point has been reached, push the wheel hard forward to take up any slack, then tighten the spindle and nut. Re-check the chain tension and the wheel alignment, before the final tightening of the spindle and nut.

4   Application of engine oil from time to time will serve as a satisfactory form of lubrication, but it is advisable to remove the chain every 2000 miles (unless it is enclosed within a chaincase, in which case every 5000 miles should suffice) and clean it in a bath of paraffin before immersing it in a special chain lubricant such as Linklyfe or Chainguard. These latter types of lubricant achieve better and more lasting penetration of the chain links and rollers and are less likely to be thrown off when the chain is in motion.

5   To check whether the chain is due for replacement, lay it lengthwise in a straight line and compress it, so that all play is taken up. Anchor one end and then pull on the other, to stretch the chain in the opposite direction. If the chain extends by more than ¼ inch per foot, replacement is necessary.

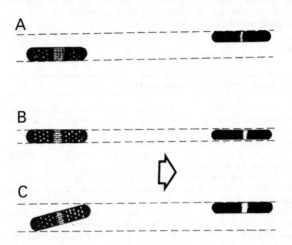

Fig. 7.7   Checking wheel alignment

B: Correct   A and C: Incorrect

6   When replacing the chain, make sure the spring link is positioned correctly, with the closed end facing the direction of travel. Reconnection is made easier if the ends of the chain are pressed into the rear wheel sprocket.

15 Rear wheel - replacement

1   The rear wheel is replaced in the frame by reversing the dismantling procedure described in Section 8 of this Chapter.

2   Make sure the distance piece is fitted between the hub and the right hand side of the swinging arm fork and that the speedometer drive gearbox has located correctly with the drive slots in the hub centre, before the wheel spindle is inserted. Check that the brake torque stop has engaged with the lug on the left hand fork tube.

3   If the brake drum has been removed from the rear wheel, make sure the three sleeve nuts are tightened fully after it has been replaced. If the nuts work loose, the studs with which they engage will be subjected to a shear stress. It is advisable to check the tightness of these nuts periodically, even if the wheel has not been removed from the frame.

16 Wheel balance

1   On any high performance machine it is important that the front wheel is balanced, to offset the weight of the tyre valve. If this precaution is not observed, the out-of-balance wheel will produce an unpleasant hammering that is felt through the handlebars at speeds from approximately 50 mph upwards.

2   To balance the front wheel, place the machine on the centre stand so that the front wheel is well clear of the ground and check that it will revolve quite freely, without the brake shoes rubbing. In the unbalanced state, it will be found that the wheel always comes to rest in the same position, with the tyre valve in the six o'clock position. Add balance weights to the spokes diametrically opposite the tyre valve until the tyre valve is counterbalanced, then spin the wheel to check that it will come to rest in a random position on each occasion. Add or subtract weight until perfect balance is achieved.

3   Only the front wheel requires attention. There is little point in balancing the rear wheel (unless both wheels are completely interchangeable) because it will have little noticeable effect on

the handling of the machine.

4   Balance weights of various sizes which will fit around the spoke nipples are available from Norton Villiers. If difficulty is experienced in obtaining them, lead wire or even strip solder can be used as an alternative, kept in place with insulating tape.

17 Tyres - removal and replacement

1   At some time or other the need will arise to remove and replace the tyres, either as the result of a puncture or because replacements are necessary to offset wear. To the inexperienced, tyre changing represents a formidable task yet if a few simple rules are observed and the technique learned, the whole operation is surprisingly simple.

2   To remove the tyre from either wheel, first detach the wheel from the machine by following the procedure in Chapters 6.2 or 7.8, depending on whether the front or the rear wheel is involved. Deflate the tyre by removing the valve insert and when it is fully deflated, push the bead from the tyre away from the wheel rim on both sides so that the bead enters the centre well of the rim. Remove the locking cap and push the tyre valve into the tyre itself.

3   Insert a tyre lever close to the valve and lever the edge of the tyre over the outside of the wheel rim. Very little force should be necessary; if resistance is encountered it is probably due to the fact that the tyre beads have not entered the well of the wheel rim all the way round the tyre.

4   Once the tyre has been edged over the wheel rim, it is easy to work around the wheel rim so that the tyre is completely free on one side. At this stage, the inner tube can be removed.

5   Working from the other side of the wheel, ease the other edge of the tyre over the outside of the wheel rim which is furthest away. Continue to work around the rim until the tyre is free completely from the rim.

6   If a puncture has necessitated the removal of the tyre, re-inflate the inner tube and immerse it in a bowl of water to trace the source of the leak. Mark its position and deflate the tube. Dry the tube and clean the area around the puncture with a petrol soaked rag. When the surface has dried, apply rubber solution and allow this to dry before removing the backing from the patch and applying the patch to the surface.

7   It is best to use a patch of the self-vulcanising type, which will form a very permanent repair. Note that it may be necessary to remove a protective covering from the top surface of the patch, after it has sealed in position. Inner tubes made from synthetic rubber may require a special type of patch and adhesive, if a satisfactory bond is to be achieved.

8   Before replacing the tyre, check the inside to make sure the agent which caused the puncture is not trapped. Check the outside of the tyre, particularly the tread area, to make sure

nothing is trapped which may cause a further puncture.

9   If the inner tube has been patched on a number of past occasions, or if there is a tear or large hole, it is preferable to discard it and fit a replacement. Sudden deflation may cause an accident, particularly if it occurs with the front wheel.

10 To replace the inner tube, inflate the inner tube sufficiently for it to assume a circular shape but only just. Then push it into the tyre so that it is enclosed completely. Lay the tyre on the wheel at an angle and insert the valve through the rim tape and the hole in the wheel rim. Attach the locking cap on the first few threads, sufficient to hold the valve captive in its correct location.

11 Starting at the point furthest from the valve, push the tyre bead over the edge of the wheel rim until it is located in the central well. Continue to work around the tyre in this fashion until the whole of one side of the tyre is on the rim. It may be necessary to use a tyre lever during the final stages.

12 Make sure there is no pull on the tyre valve and again commencing with the area furthest from the valve, ease the other bead of the tyre over the edge of the rim. Finish with the area close to the valve, pushing the valve up into the tyre until the locking cap touches the rim. This will ensure the inner tube is not trapped when the last section of the bead is edged over the rim with a tyre lever.

13 Check that the inner tube is not trapped at any point. Re-inflate the inner tube, and check that the tyre is seating correctly around the wheel rim. There should be a thin rib moulded around the wall of the tyre on both sides, which should be equidistant from the wheel rim at all points. If the tyre is unevenly located on the rim, try bouncing the wheel when the tyre is at the recommended pressure. It is probable that one of the beads has not pulled clear of the centre well.

14 Always run the tyres at the recommended pressures and never under or over-inflate. The correct pressures for solo use are given in the Specifications Section of this Chapter. If a pillion passenger is carried, increase the rear tyre pressure as recommended.

15 Tyre replacement is aided by dusting the side walls, particularly in the vicinity of the beads, with a liberal coating of French chalk. Washing-up liquid can also be used to good effect, but this has the disadvantage of causing the inner surfaces of the wheel rim to rust.

16 Never replace the inner tube and tyre without the rim tape in position. If this precaution is overlooked there is good chance of the ends of the spoke nipples chafing the inner tube and causing a crop of punctures.

17 Never fit a tyre which has a damaged tread or side walls. Apart from the legal aspects, there is a very great risk of a blow-out, which can have serious consequences on any two-wheel vehicle.

18 Tyre valves rarely give trouble, but it is always advisable to check whether the valve itself is leaking before removing the tyre. Do not forget to fit the dust cap which forms an effective second seal.

Tyre removal
A   *Deflate inner tube and insert lever in close proximity to tyre valve*
B   *Use two levers to work bead over the edge of rim*
C   *When first bead is clear, remove tyre as shown*

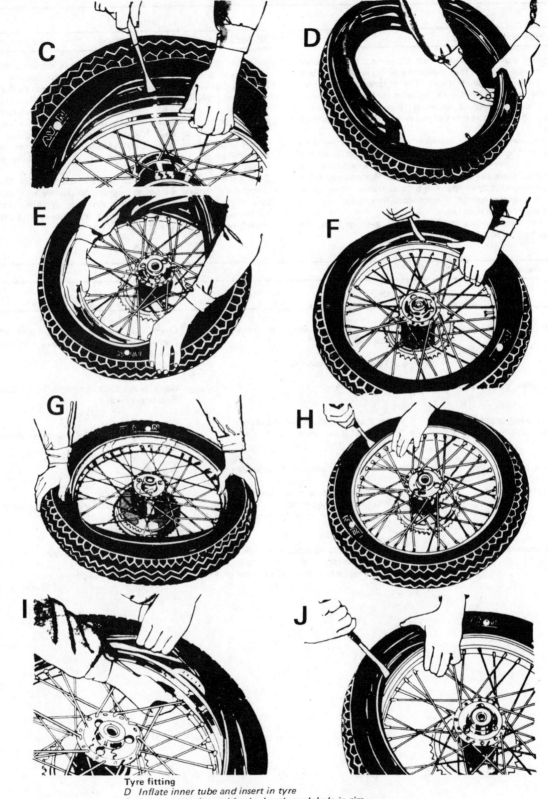

Tyre fitting
D Inflate inner tube and insert in tyre
E Lay tyre on rim and feed valve through hole in rim
F Work first bead over rim, using lever in final section
G Use similar technique for second bead. Finish at tyre valve position
H Push valve and tube up into tyre when fitting final section, to avoid trapping

Security bolts
I Fit the security bolt very loosely when one bead of the tyre is fitted.
J Then fit tyre in normal way. Tighten bolt when tyre is properly seated.

18 Tyre valve dust caps

1 Tyre valve dust caps are often left off when a tyre has been replaced, despite the fact that they serve an important two-fold function. Firstly, they prevent dirt or other foreign matter from entering the valve and causing the valve to stick open when the tyre pump is next applied. Secondly, they form an effective second seal so that in the event of the tyre valve sticking, air will not be lost.

2 Isolated cases of sudden deflation at high speeds have been traced to the omission of the dust cap. Centrifugal force has tended to lift the tyre valve off its seating and because the dust cap is missing, there has been no second seal. Racing inner tubes contain provision for this happening because the valve inserts are fitted with stronger springs, but standard inner tubes do not, hence the need for the dust cap.

3 Note that when a dust cap is fitted for the first time, the wheel may have to be rebalanced.

19 Security bolt - function and fitting

1 If the drive from a high-powered engine is applied suddenly to the rear wheel of a motor cycle, wheel spin will occur with an initial tendency for the wheel rim to creep in relation to the tyre and inner tube. Under these circumstances there is risk of the valve being torn from the inner tube, causing the tyre to deflate rapidly, unless movement between the rim and tyre can be restrained in some way. A security bolt fulfills this role in a simple and effective manner, by clamping the bead of the tyre to the well of the wheel rim so that any such movement is no longer possible.

2 A security bolt is fitted to the rear wheel of all Norton Commando models. Before attempting to remove or replace a tyre, it must be slackened off completely so that the clamping action is released. The accompanying tyre fitting illustrations show how the security bolt is fitted and secured.

20 Fault diagnosis

Symptom	Cause	Remedy
Handlebars oscillate at low speeds	Buckle or flat in wheel rim, most probably front wheel	Check rim alignment by spinning wheel. Correct by retensioning spokes or rebuilding on new rim.
	Tyre not straight on rim	Check tyre alignment.
Machine lacks power and accelerates poorly	Brakes binding	Warm brake drum provides best evidence. Re-adjust brakes.
Brakes grab when applied gently	Ends of brake shoes not chamfered	Chamfer with file.
	Elliptical brake drum	Lightly skim in lathe (specialist attention required).
Front brake feels spongy	Air in hydraulic system (disc brake only)	Bleed brake.
Brake pull-off sluggish	Brake cam binding in housing	Free and grease.
	Weak brake shoe springs	Renew if springs have not become displaced.
	Sticking pistons in brake caliper (front disc brake only)	Overhaul caliper unit.
Harsh transmission	Worn or badly adjusted final drive chain	Adjust or renew as necessary.
	Hooked or badly worn sprockets	Renew as a pair.
	Loose rear wheel (Pre-1971 models only)	Check wheel retaining bolts.

Chapter 8 Electrical system

Contents

Specifications

Battery

Manufacturer	Lucas or Yuasa
Type	Lead/acid, 12 volt
Capacity	10 amp/hr *
Earth connection	Positive

* capacity will differ on 'Interpol' models or those where two 6V batteries connected in series are fitted

Alternator

Manufacturer	Lucas
Type	RM 21

Rectifier

Manufacturer	Lucas
Type	2 DS 506

Zener Diode

Manufacturer	Lucas
Type	ZD 715

Warning light assimilator

Manufacturer	Lucas
Type ... ...•...	3 AW

Flashing indicator lamp relay

Manufacturer	Lucas
Type	8 FL

Horn

Manufacturer	Lucas
Type	† 9H

† Interpol models have twin Lucas 9H alternating horns, working in conjunction with an HC 3 alternating horn control
unit and 6 RA relays

Headlamp

Manufacturer	Lucas
Type	SS700P or MCH66 (Hi-Rider model)

Bulbs

Main headlamp	45/40W or 60/55W (Quartz - halogen)
Pilot lamp	6W
Tail/stop lamp	6/21W
Indicator lamps	2W
Speedometer/tachometer illumination	2.2W
Flashing indicator lamps	21W

All bulbs are 12 volt

1 General description

All 750 cc and 850 cc Norton Commando models are fitted with a 12 volt electrical system. The circuit comprises a crankshaft-driven alternator with a six pole permanent magnet rotor which rotates within a six pole laminated iron stator coil assembly, a rectifier to convert the alternating current into direct current for charging the battery, and the battery itself. Voltage regulation is achieved by means of a Zener diode, a semi-conductor device which becomes conductive in the reverse flow direction when a predetermined voltage level is reached.

Before any electrical tests are carried out, the warning light assimilator fitted to all post-1970 models must be disconnected by removing the green/yellow lead where it joins the alternator double snap connector.

2 Crankshaft alternator - checking the output

1 As explained in Chapter 5, Section 2.1, the ammeter mounted in the top of the headlamp shell of the earlier models provides visual indication of the output from the alternator, whilst the engine is running. If no charge is indicated, even with a full lighting load, the performance of the alternator is suspect and should be investigated further. Later models do not have this facility for checking, only a red 'charge' lamp.

2 To allow tests to be carried out on the alternator and also the rectifier a suitable heavy duty one ohm resistor must be constructed. A suitable resistor can be made using about 4 feet (1220 mm) of 18 SWG (0.048 in diameter) Nichrome wire. This wire can be obtained through auto-electrical specialists. Bend the length of wire in half, and attach a probe lead and crocodile clip at the bend in the wire. Remove the battery from the machine, and connect the probe lead clip to the positive (+) terminal. Connect a second lead from the negative terminal, and connect this to one terminal of a 0-20 amp d.c. ammeter. A 0-20 volt d.c. voltmeter should be connected between the battery terminals. Finally, take a lead from the remaining terminal of the ammeter and fit a crocodile clip which can be connected to the far end of the Nichrome wire. Resistance can be calculated by dividing the voltage by the number of amps; for example 12 volts divided by 3 amps would indicate a

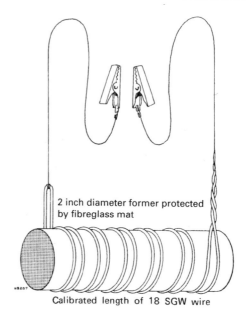

2 inch diameter former protected by fibreglass mat

Calibrated length of 18 SGW wire

Fig. 8. 2. 1 ohm resistor

resistance of 4 ohms. The resistance can be varied by moving the crocodile clip up or down the Nichrome wire. One ohm will be obtained when the number of volts equals the number of amps indicated on the meters. When the correct position for the second lead has been established, cut the doubled Nichrome wire at that point, and secure the second probe lead. To make the resistor less unwieldy, the wire can be wound in coils around a 2 in diameter former as shown in Fig. 8.2, spacing the coils so that they do not touch. Note that the resistor will become hot in use. To avoid burning, wrap the former with a Plumber's fibreglass soldering/brazing mat. This, and the Nichrome wire may be purchased from online retailers.

3 To check the output from the alternator, disconnect the

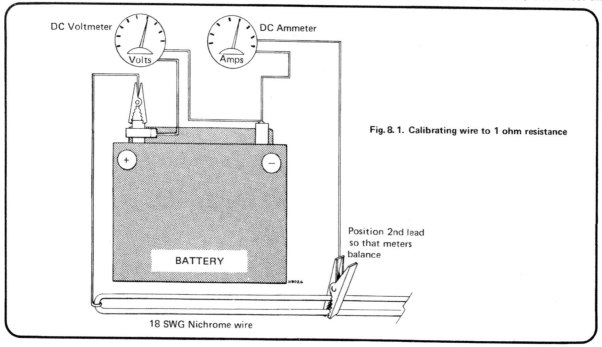

DC Voltmeter

Volts

DC Ammeter

Amps

BATTERY

Position 2nd lead so that meters balance

18 SWG Nichrome wire

Fig. 8. 1. Calibrating wire to 1 ohm resistance

alternator leads which emerge from the rear of the primary chaincase by means of the snap connectors provided. Connect an AC voltmeter with a 1 ohm resistance in parallel between the white/green and green/yellow leads, then start the engine and run it at approximately 3000 rpm. If the alternator is functioning correctly, an output of 9 volts should be recorded as a minimum.

4   Stop the engine, then connect one lead from the voltmeter and resistance to earth, whilst the other lead is applied to the white/green and green/yellow wires in turn. The engine should be started and run at 3000 rpm after each connection is made. During both these checks the voltmeter should not show a reading.

5   If the tests produce results that do not correspond with the expectations listed in the preceding paragraphs, fit a substitute alternator and re-check. If the alternator proves to be defective, a new replacement can be obtained at reduced cost, through the Lucas Service Exchange Scheme.

3 Battery - examination and maintenance

1   All models have a 12 volt, 10 amp/hr battery with a translucent plastic case, except the Interpol model, which has two 6 volt batteries mounted in series.

2   Battery maintenance is limited to keeping the electrolyte level just above the plates and separators, as denoted by the level line marked on the case. The level of the electrolyte is readily visible, due to the translucent nature of the case. Do not overfill and make sure the vent pipe is attached, so that it will discharge away from any parts of the machine liable to suffer damage from corrosion.

3   Unless acid is spilt, which may occur if the machine falls over, use only distilled water for topping up purposes, until the correct level is restored. If acid is spilt on any part of the machine, it should be neutralised immediately with an alkali such as washing soda or baking powder, and washed away with plenty of water. This will prevent corrosion from taking place. Top up in this instance with sulphuric acid of the correct specific gravity (1.260 - 1.280).

4   It is seldom practicable to repair a cracked battery case because the acid which is already seeping through the crack will prevent the formation of an effective seal, no matter what sealing compound is used. It is always best to replace a cracked battery, especially in view of the risk of corrosion from the acid leakage.

5   Make sure the battery is clamped securely. A loose battery will vibrate and its working life will be greatly shortened, due to the paste being shaken out of the plates.

6   Ensure the battery connections are kept clean and tight. If corrosion occurs, clean the parts concerned by immersing them in a solution of baking powder to neutralise the acid. When reconnection is made, apply a light smearing of petroleum jelly such as vaseline, to prevent corrosion recurring.

7   All Norton Commando models have a positive earth system.

4 Silicon rectifier - function and testing

1   The function of the silicon rectifier is to convert the AC current from the alternator into DC current for charging the battery. The rectifier is of the full wave type and does not require any maintenance.

2   The rectifier is located close to the battery, in the compartment below the dualseat. It is located in this position so that it is not directly exposed to water or to accidental damage.

2   One of the most frequent causes of rectifier failure is caused by the inadvertent connection of the battery in reverse, which initiates a reverse flow of current through the rectifier. It is not practicable to repair a damaged rectifier; renewal is the only satisfactory solution.

3   The usual indication of rectifier failure is the inability of

the charging system to meet the machine's electrical demands. To check whether the rectifier is functioning correctly, disconnect the blue/brown lead from the centre terminal of the rectifier and lodge it in a position where it cannot short to earth. Connect the negative lead of a DC voltmeter to the centre terminal of the rectifier, and the positive lead to a convenient earth point. A one ohm resistor of the type described in Section 2, paragraph 2 should be connected in parallel with the voltmeter. If the rectifier is functioning correctly, a reading of 7½ volts at 3000 rpm should be obtained when the engine is run.

4   If for any reason the rectifier has to be removed, care should be taken to ensure the plates are not twisted in relation to one another, or even scratched or bent. It is only too easy to damage the internal connections and render the rectifier unfit for further service. The centre bolt must ALWAYS be held firmly with a spanner, whilst the mounting nut is removed or replaced.

5 Fuse - location and replacement

1   A 35 amp fuse is fitted into the battery negative lead, close to the battery carrier. It is retained within a nylon holder and is of the replaceable type. To release the fuse, push both halves of the fuse holder body together and twist.

2   If a fuse blows, it should be replaced, after checking to ensure that no obvious short circuit has occurred. If the second fuse blows shortly afterwards the electrical circuit must be checked thoroughly, to trace and eliminate the fault.

3   Always carry at least one spare fuse. If a situation arises where a fuse blows whilst the machine is in use and no spare is available, a 'get you home' remedy is to remove the defective fuse and wrap it in silver paper before replacing it in the fuse holder. The silver paper will restore electrical continuity by bridging the broken fuse wire. This expedient should NEVER be used if there is evidence of a short circuit or other major electrical fault, otherwise more serious damage will be caused. Replace the blown fuse at the earliest possible opportunity, to restore full circuit protection.

4   For similar reasons, do not fit a fuse which has a higher rating than the original, or it may no longer form the weakest link in the circuit protection chain.

6 Zener diode - function, location and testing

1   The Zener diode performs the voltage regulating function accepting excess current from the alternator which is not required for battery charging purposes and converting it into heat which is dissipated by the mounting (heat sink) to which the diode is attached. It follows, therefore, that the Zener diode must be located so that it is rigidly attached to a good heat conductive base which will dissipate the heat by its location in a continuous air stream. The Zener diode is mounted on the inside of the right hand alloy footrest support plate.

2   Contact between the base of the Zener diode and the heat sink must be clean and corrosion-free, otherwise the heat path will be interrupted. The tightening torque is critical. It must not exceed 0.28 kg m or there is risk of the mounting stud shearing.

3   To check the operation of the diode, connect a DC ammeter between the snap connector and feed cable, positive lead to the diode terminal and negative lead to the feed cable. Connect a DC voltmeter between the diode terminal and earth, negative lead to the terminal and the positive lead to earth. Ensure the lights are switched off and that there is no other electrical load applied, then start the engine and gradually increase the speed. The Zener diode must be replaced if the ammeter shows a reading before the voltmeter records 12.75 volts, or if the voltmeter shows more than 15.5 volts before the ammeter reading has reached 2 amps. If the Zener diode is faulty, it must be renewed. A repair is not possible.

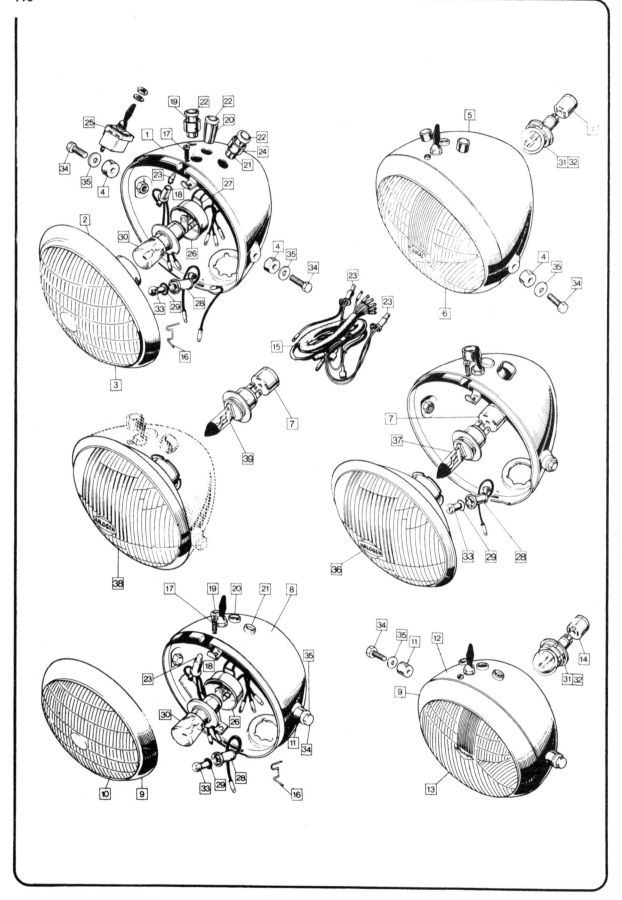

6.1 Zener diode is attached to rear of right hand alloy footrest support plate

7.1 Reflector unit is released by screw in top of headlamp shell

4   Throughout the test, the battery must be in a fully charged condition, or a completely false reading may result. If the state of charge is low, fit a fully-charged battery for the duration of the test.

7 Headlamp - renewing bulbs and adjusting beam height

1   The headlamp is fitted with a sealed beam unit of the quartz iodine type, of which the main bulb is of the pre-focus type. The complete unit is secured to the headlamp rim by wire clips; access is gained by slackening the screw in the top of the headlamp shell (close to the rim) and then lifting the rim away with the sealed beam unit attached. A push-fit multi-point connector forms the detachable electrical connections with the main headlamp bulb, located within the sealed unit. If either the main or dipped filaments fail, the main bulb must be renewed. The bulb holder is released by pressing inwards and turning to the left.

2   The pilot bulb and bulb holder are attached to the reflector of the sealed beam unit. The pilot bulb has a bayonet fitting and is easily renewed when the bulb holder is pulled out of the reflector.

3   Beam height is adjusted by slackening the two bolts which retain the headlamp shell to the forks and tilting the headlamp either upwards or downwards. Adjustments should always be made with the rider seated normally.

4   UK lighting regulations stipulate that the lighting system must be arranged so that the light does not dazzle a person standing in the same horizontal plane as the vehicle, at a distance greater than 25 yards from the lamp, whose eye level is not less than 3 feet 6 inches above that plane. It is easy to approximate this setting by placing the machine 25 yards away from a wall, on a level road, and setting the beam height so that it is concentrated at the same height as the distance from the centre of the headlamp to the ground. The rider must be seated normally during this operation, and the pillion passenger, if one is carried regularly.

5   If the sealed beam unit is broken, it can be removed from the headlamp rim by detaching the wire retaining clips, after the rim has been removed from the front of the headlamp.

6   A pilot bulb of 12 volt, 6 watt rating is fitted. The fitting is the miniature bayonet cap type.

7   Variations in headlamp occur, depending on the model to which the lighting system is fitted and its ultimate destination. Some of the early models and also those destined for France use a main headlamp bulb of either the Lucas Pre-Focus or Continental type, and may or may not have an ammeter fitted in the top of the headlamp shell. The Hi-Rider model is fitted with the very distinctive Lucas MCH66 headlamp, which has a flat back and is shallow in depth. The general internal arrangement is, however, similar.

FIG. 8.3. HEADLAMPS

1 Lucas SS700P headlamp	21 Green warning lamp body
2 Rim	22 Sealing ring for lamp body 3 off
3 Reflector unit	23 Warning lamp bulb 3 off
4 Spacer	24 Warning lamp shield 3 off
5 Lucas SS700P headlamp for Continent/France	25 Lighting switch
6 Reflector unit for Continent/France	26 Main bulb holder
7 Adaptor	27 Terminal sleeve
8 Lucas MCH66 headlamp for Hi-Rider model	28 Pilot bulb holder
9 Rim	29 Pilot bulb seating ring
10 Reflector unit	30 Main bulb, 45/40W
11 Spacer 2 off	31 Continental main bulb, 45/40W
12 Lucas MCH66 headlamp for Continent/France	32 French main bulb, 45/40W
13 Reflector unit for Continent/France	33 Pilot bulb
14 Adaptor	34 Headlamp mounting bolt 2 off
15 Headlamp wiring harness	35 Headlamp mounting washer 2 off
16 Reflector unit clips 6 off	36 Quartz-halogen light unit
17 Rim fixing screw	37 Quartz-halogen bulb
18 Rim fixing plate	38 Continental light unit (Quartz-halogen)
19 Amber warning lamp body	39 Continental bulb (Quartz-halogen)
20 Red warning lamp body	

8 Indicator lamps - renewal

1   The headlamp shell will have at least one indicator lamp mounted close to the headlamp rim. The usual arrangement is three lamps, as follows:

Amber	Flashing indicator repeater
Red	Charge indicator (in lieu of ammeter)
Green	Headlamp main beam indicator

2   The bulb holders pull out from the inside of the headlamp shell, making it necessary to detach the front of the headlamp unit in order to gain access. The bulbs are of the miniature bayonet fitting type, rated at 12 volts, 2 watts.

9 Lighting switch

The lighting switch fitted to the pre-1971 models is located in the top of the headlamp shell and has three positions: off, pilot light and main beam. Models manufactured from 1971 onwards have only a two position switch mounted in the same position. The lighting selection is controlled by a combined lighting/ignition switch, as described in the following section. The headlamp switch changes over from pilot to main beam when the 'lights on' position has been selected on the combined lighting/ignition switch.

10 Ignition and master switches

1   Pre-1971 models have an ignition switch mounted in front of the left hand accessory cover, on the frame tube which curves downwards from the underside of the petrol tank. The switch is of the two-position variety, having an on and an off position. It is actuated by a key which is retained in the lock whilst the ignition is switched on. The same key fits the lock on the steering head of the machine.
2   If the switch malfunctions, it must be renewed; it is not practicable to effect a repair. Note that if the switch is renewed, it will be necessary to retain the original key for the steering head lock.
3   To detach the switch, peel back the switch cover and displace the tumbler assembly by depressing the spring-loaded plunger through the hole in the switch body. If the switch body is shaken whilst the plunger is depressed, the plunger should be ejected. Note the plunger has the key number stamped on its side, a useful means of identifying the correct key if the original is lost.
4   Models made from 1971 onwards have a four-position ignition and lighting switch, also actuated by the ignition key. The key can be removed only when the switch is in the off position, or when the parking lights only are switched on.

11 Warning light assimilator

1   From 1971 onwards, a warning light assimilator was included with the ignition coil assembly, to give visual indication by means of a warning lamp in the headlamp shell which the alternator is charging. When the ignition is switched on, and before the engine is started, the lamp is lit by current from the battery. When the engine starts and the alternator output reaches the 6 volt level, contacts within the assimilator unit open, which take the warning lamp out of circuit.
2   If the warning lamp fails to light when the ignition is switched on, suspect a faulty bulb or wiring. A check can be made by detaching the white/brown lead from the WL terminal of the assimilator unit and temporarily connecting it to earth. If the indicator lamp lights up, the assimilator unit is faulty or the alternator has no output. Proceed with the alternator checks, as described in Section 2 of this Chapter.

12 Tail and stop lamps - renewing bulbs

1   Removal of the plastic lens cover, which is retained by two long screws, will reveal the bulb holder which holds the combined tail and stop lamp. The bulb is of the bayonet fitting type, but has offset pins to prevent accidental inversion. The tail lamp filament is rated at 6 watts and the stop lamp filament at 21 watts.
2   The stop lamp switch actuated by the rear brake is attached to the brake pedal. It is adjusted as described in Chapter 7, Section 13.2. Disc brake models have an additional stop lamp switch incorporated in the master cylinder unit. This switch is not adjustable. Drum brake models have a sealed compression switch which forms an integral part of the front brake cable. This switch too cannot be adjusted.

13 Handlebar switch clusters

1   Switch clusters incorporated in light alloy castings attached to the handlebar ends are a feature of all 1971 and later models. Although not sealed units, access is difficult if a switch malfunctions. This type of fault is best remedied by renewing the part concerned.
2   If desired, the switch clusters can be reversed, to suit rider

12.1 Tail/stop lamp bulb is of the offset pin type

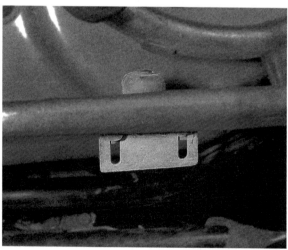

12.2 Stop lamp switch is attached to rear brake pedal. Is adjustable for height convenience.

14 Flashing indicator lamps and relay

1 Flashing indicator lamps, to indicate the direction of turn, are fitted as standard equipment to all late models. They are mounted on hollow 'stalks' through which the electrical leads pass. The flasher unit is located in close proximity to the battery, beneath the seat.

2 Access to each flashing indicator bulb is obtained by removing the plastic lens cover, secured by two screws. The bulbs are of the bayonet fitting type and are rated at 12 volts.

3 A series of audible clicks will be heard if the flasher unit is functioning correctly. If the unit malfunctions, the usual symptom is one initial flash before the unit goes dead. It will be necessary to renew the flasher unit if the fault cannot be attributed to a burnt-out indicator bulb or a blown fuse. Handle the new unit with care, as it is easily damaged if dropped or subjected to shock. The unit is located under the seat, bolted to the battery compartment, above the rear mudguard.

15 Speedometer and tachometer bulbs

The speedometer and tachometer heads each have an internal bulb to illuminate the dial during the hours of darkness. The bulb holder is a push fit in the underside of each instrument. A bulb rated at 12 volts, of the miniature bayonet fitting type, is fitted to each.

16 Horn adjustment

1 The horn is suspended from the underside of the battery carrier, facing forwards across the frame, in line with the top of the alloy footrest support plates. Adjustment will be necessary from time to time, to take up wear in the moving parts. Late models have twin horns with a dissimilar note.

2 To adjust the horn, press the horn button and turn the adjusting screw (in the eleven o'clock position in the back of the horn) anticlockwise until the horn just fails to sound. Release the horn button and turn the adjusting screw clockwise one notch at a time until horn performance is restored when the horn button is pressed. The amount of adjustment required usually varies from one quarter to three-quarters of a turn.

3 If the horn fails to respond to this treatment, further attention is required by an electrical specialist.

4 The Police Interpol is supplied with an alternating horn set to give the special two-tone note associated with this type of vehicle. The system comprises twin horns, a transistorised multi-vibrator to provide the alternating notes, two relays, a switch and the various connectors.

5 The twin horns, both of the Lucas 9H type, must be mounted rigidly on a solid member of the machine, with no loosely-mounted components in the vicinity. If this precaution is not observed, the tone will be adversely affected. If possible, the horns should be tilted slightly downwards to permit any water which may enter the horns to drain away. They should be mounted facing in a forward direction.

6 The horns will require adjustment periodically, to compensate for wear of the contact breaker mechanism. Adjustment is effected as follows: Connect a voltmeter across the terminals of the horn, to monitor the voltage whilst adjustment is carried out. Connect an ammeter (0 - 25 amp range) between terminals C1 and C2 of the relay in the circuit of the horn being adjusted. The adjustment screw (located in the eleven o'clock position in the back of the horn) should be turned whilst the horn is operated, until the ammeter reads from 3.0 and 3.5 amps in the case of the high note horn and from 3.0 to 4.0 amps in the case of the low note horn. Turning the screw clockwise will reduce the ammeter reading. The setting voltage throughout should be 13 volts, but after adjustment the horns should be capable of producing a good clear note over the 11 to 15 volt range.

7 Do not disturb the central slotted screw and locknut during the adjusting operation, or sound the horn when the contacts are out of adjustment. In view of the heavier electrical loading which may result during adjustment, it is advisable to short out the fuse until adjustment is complete.

8 The standard 6RA continuously rated relays should be mounted so that the cables carrying the horn current are as short as possible. This will prevent unnecessary electrical losses. If the horns fail to sound when the switch is operated, the relay unit should be checked as follows: Connect the positive lead of a voltmeter to the positive terminal of the relay unit, and connect the cable which is displaced to the negative lead of the voltmeter. The operating voltage should be indicated when the switch is actuated; if the reading is low or non-existant, the fault lies in the supply to the relay unit. Check each relay in turn by removing the leads from one 'R' terminal and the positive terminal, then join the two leads together. Operate the switch and if the horn sounds, repeat this procedure, connecting the other 'R' terminal to the positive lead. If both horns and relays operate, the transistorised control unit is at fault and must be renewed. If one of the horns does not sound, earth the leads connected to the C1 terminal of the relay involved. If the horn operates satisfactorily, the fault is in the relay, necessitating renewal of the relay unit.

9 The transistorised multi-vibrator is pre-set during manufacture and it is not practicable to change the pulse frequency. The unit must be renewed if a fault develops. It should be mounted so that it is not subjected to a temperature exceeding 50ºC and care should be taken to ensure the electrical polarity is not reversed at the connections.

17 Wiring - layout and examination

1 The wiring is colour-coded and will correspond with the accompanying wiring diagrams.

2 Visual inspection will show whether any breaks or frayed outer coverings are giving rise to short circuits. Another source of trouble may be the snap connectors, particularly where the connector has not been pushed home fully in the outer casing. Early models are especially prone to wiring faults because the rubber-covered cables used at that period deteriorate as time progresses.

3 Intermittent short circuits can sometimes be traced to a chafed wire which passes through a frame member. Avoid tight bends in the wire or situations where the wire can become trapped or stretched between casings.

18 Fault diagnosis

Symptom	Cause	Remedy
Complete electrical failure	Blown fuse	Check wiring and electrical components for short circuit before fitting new 15 amp fuse.
	Isolated battery	Check battery connections, also whether connections show signs of corrosion.
Dim lights, horn inoperative	Discharged battery	Recharge battery with battery charger and check whether alternator is giving correct output.
Constantly 'blowing' bulbs	Vibration, poor earth connection	Check whether bulb holders are secured correctly. Check earth return or connections to frame.

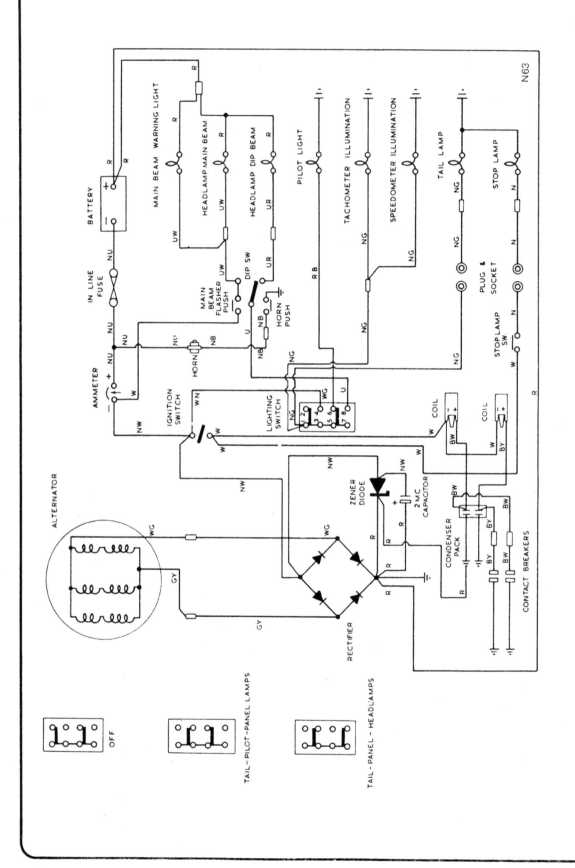

Fig. 8.4. Pre-1971 wiring diagram, all models

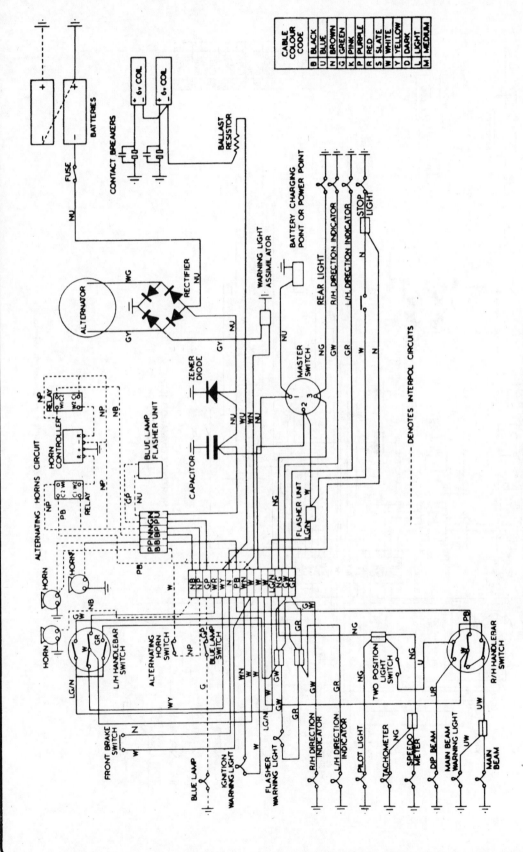

Fig. 8.5. 1971 wiring diagram, all models

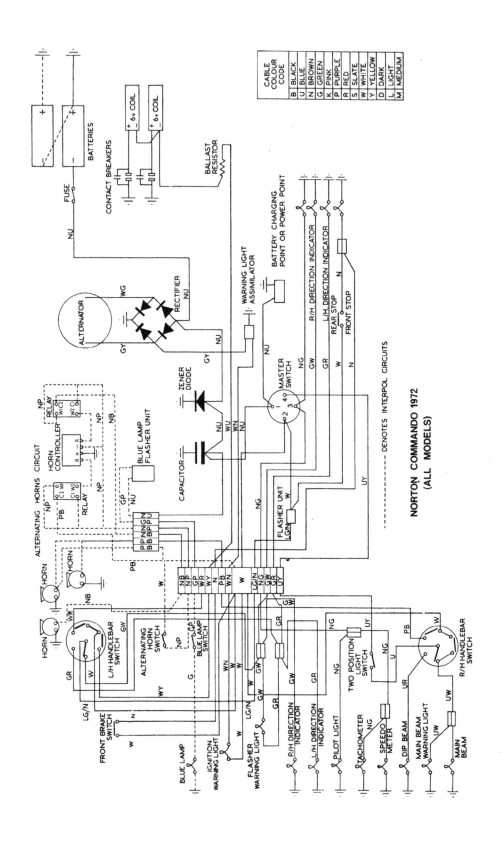

Fig. 8.6. 1972 onwards wiring diagram, all models

NORTON COMMANDO 1972
(ALL MODELS)

CABLE COLOUR CODE	
B	BLACK
U	BLUE
N	BROWN
G	GREEN
K	PINK
P	PURPLE
R	RED
S	SLATE
W	WHITE
Y	YELLOW
D	DARK
L	LIGHT
M	MEDIUM

- - - - - DENOTES INTERPOL CIRCUITS

Chapter 9 The Mark 3 electric start model

Contents

1 Introduction and major modifications

Noticeable external differences found on the Mark 3 Electric Start models are; the rear wheel which is disc braked, new electrical controls , the addition of an electric starter, left-hand gear change conversion and the relocation of the front disc brake assembly on the left-hand fork leg.

Internally there are other differences which make some of the internal components non-interchangeable with those used on previous models. A major change occurs on the primary drive side and relates to the addition of an electric starter and its essential components. Also of importance is the new Isolastic engine mountings, which are adjustable and no longer require re shimming - originally a time consuming maintenance task. The electrical system has been altered considerably to incorporate new switchgear, a higher output alternator and the electric starter.

2 Engine and frame components - dismantling, examination and reassembly - general

1 Although the Mark 3 model is very similar in many respects to the models described in the preceeding Chapters, reference should always be made to this Chapter first, where the major differences and the modified procedures necessary are covered in detail. Where no information is given, it can be assumed that the procedures are identical to those given for the 850 model in the earlier Chapters.

2 Routine maintenance procedures follow the same schedule, already described in an earlier section of this manual.

3 This Chapter has been grouped together in the same Chapter and Section order as the main text of the manual. Wherever possible, the same Section headings have been used to facilitate easy cross reference to the main text

3 Dismantling the engine - removing the cylinder head

1 Most models are fitted with a downswept exhaust system. This is removed by bending back the tab washer on each exhaust locking ring to free the exhaust pipe from the cylinder head. Norton Villiers service tool 063968 is recommended for this purpose as the rings are locked tight. If the service tool is not available, the rings can be slackened by careful application of a

1975 Mark 3 850 cc electric start model

flat nosed punch and hammer. The 850 cc models have a balance pipe, the clips of which must be slackened.

2   To release the silencers, unscrew the nuts which secure the mounting plates to the two rubber mountings. The exhaust pipes and silencers can then be lifted away as a complete unit.

3   A somewhat similar procedure is recommended for the unswept exhaust system fitted to the 'S' models. In this instance, the silencer has only one point of attachment, on a bracket adjacent to the rear suspension unit.

4   Note there is a copper/asbestos sealing ring in each exhaust port, which should be removed and discarded. It is customary to fit new replacements when the exhaust system is eventually refitted in order to preserve a leaktight joint.

5   Remove both carburettors complete with spacers by unscrewing the four socket screws which retain the assembly to the cylinder head. Access is made easier by using a short socket key in order to provide clearance with the frame tube. Disengage the air cleaner hoses from each carburettor intake and lift the carburettors away. If desired, the carburettors can be separated by disconnecting the balance pipe which joins them and by removing each carburettor top, retained by two crosshead screws. This will enable the slide and needle assembly, complete with control cables, to be lifted out of each mixing chamber.

6   Whichever method is used for the removal of the carburettor, ensure the parts involved are taped out of harms way. They are very easily damaged if mishandled. Lift away the air cleaner assembly by withdrawing the two long bolts which hold the case together.

7   Remove the cylinder head steady. This takes the form of two short plates which interconnect a plate bolted to the cylinder head with two rubber mountings, one on each side of the tube below the main frame tube. Remove the nuts from the rubber mounting in order to prevent the mountings from rotating, and remove the side plates. Slacken and remove the 3 Allen screws retaining the steady to the cylinder head. This also releases the tension on the suspensory spring device, which preferably should be left intact, otherwise the device will have to be reset (see Section 12 of this Chapter). Detach the spring stirrup from the coil bracket noting in which notch the loop of the spring was located.

8   Unbolt the twin ignition coil assembly from its mounting above the cylinder head and disconnect both caps from the spark plugs. The coil assembly need not be removed completely; it can be tied to the handlebars so that it does not impede removal of the cylinder head. Only the electrical connections from the coils to the contact breakers need be removed.

9   Remove both spark plugs and remove the rocker oil feed pipes from both sides of the cylinder head casting. Special care is needed when unscrewing the banjo union bolts to prevent the thin copper pipe from twisting or necking. Detach the copper sealing washers from either side of each union but leave the feed pipe attached to the timing cover.

10   Slacken and remove nine of the cylinder head retaining bolts, leaving only the front centre bolt in position. This and the two bolts, one on either side, are recessed into the cylinder head and will require a slim socket or box spanner for their release. Two nuts are found on the underside of the cylinder head, at the front, and another single nut on the underside, at the rear. The four remaining bolts are easily accessible from the top of the cylinder head, making a total of ten bolts, 850 cc models include four through bolts in the cylinder barrel.

11   When the front cylinder head bolt is removed last of all, the cylinder head will tilt a little against the spring pressure of the valve which is open. This will aid breaking the cylinder head joint.

12   Before the cylinder head can be lifted away, it is necessary to feed each of the four pushrods into the cylinder head as far as possible, after detaching them from the ends of the rocker arms. This can be accomplished by tilting the cylinder head towards the rear, whilst holding the cylinder head with one hand and the pushrods with the other. Do not use force and make sure the pushrods are clear of the cylinder barrel as the head is being

3.7 Do not undo the locknut of the suspensory spring device

4.5 Remove the primary chaincase outer cover

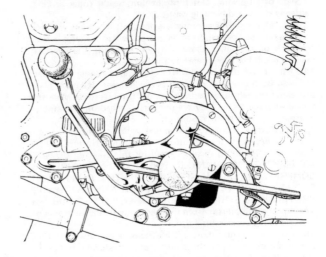

Fig. 9.1 Method of wedging rear brake

removed. Failure to observe this precaution may cause damage to the light alloy pushrods, necessitating their renewal.

13 The cylinder head gasket will adhere to either the cylinder barrel or cylinder head and should not be re-used unless it is completely undamaged.

4 Dismantling the engine - removing the alternator, clutch primary chaincase and electric starter

1   Engage first gear. Wedge the rear brake on with a steel bar (see illustration 9.1) to prevent the various components from turning whilst being dismantled. Alternatively, if the cylinder head and block has been removed, lock the engine with a steel bar placed through the small ends of the connecting rods. Do not let the bar bear directly onto the crankcase ; use two pieces of wood to protect the surface.

2   Disconnect the electric cable to the starter motor. Remove the engine earth lead at the rear left-hand side of the crankcase.

3   Remove the left- hand footrest by withdrawing the two bolts. Note the red earth wire from the zener diode.

4   Place a large drain tray underneath the primary chaincase and remove the drain plug, located below the alternator bulge.

5   Remove the eleven retaining screws and gently knock the cover complete with gear lever off its locating dowels, using a soft faced hammer.

6   Withdraw the gear lever crossover shaft complete with pinion from its splined connector in the rubber sleeve.

7   Remove the alternator rotor nut using a ½ in. Whitworth socket. Do not misplace the washer which seats below the nut.

8   Remove the alternator stator which is secured by three nuts and washers. Draw the stator assembly off the studs, after detaching the lead wire at the snap connectors in the vicinity of the air cleaner housing. The lead wire will pull through the small rubber grommet in the centre of the rear chaincase.

9   The alternator rotor is keyed onto the crankshaft and has a parallel fit. In consequence, it is not difficult to remove. Light pressure with a pair of tyre levers positioned at the rear of the rotor should provide sufficient leverage. Remove the rotor key, the packing collar and any shims on the crankshaft. Detach the three spacers from the stator mounting studs and place them in a safe position for reassembly.

10 Slacken the clutch pushrod adjuster nut in the centre of the clutch assembly and remove both the adjuster and the nut.

11 Knock back the tabs on the four lock washers of the alternator outrigger plate. Remove the nuts and washers and lift off the plate.

12 Slide off the sprag clutch mechanism, needle roller bearing, hardened sleeve and backing washer.

13 Pull out the starter intermediate gearshaft, complete with the backfire overload device, from its splined hole in the large gear at the back of the inner cover. Do not dismantle the overload device since it has been preset by the factory.

14 First remove the small nut from the primary chain tensioner; followed by the two ½ in. AF nuts which release the outer plate and allow the tensioner to be lifted off from its studs. Mark the tensioner plungers so they are not interchanged on reassembly; alternatively, wire them in position.

15 The clutch cannot be dismantled without the compressor which is necessary for the diaphragm spring. Norton Villiers service tool 060999 is recommended for this purpose. DO NOT ATTEMPT TO DISMANTLE THE CLUTCH WITHOUT A COMPRESSOR. IF THE TENSION OF THE DIAPHRAGM SPRING IS RELEASED SUDDENLY WITHOUT PROTECTION, SERIOUS INJURY MAY RESULT. Do not unscrew the compressor from the diaphragm or the latter will be released with considerable force.

16 Screw the centre bolt of the compressor into the hole previously occupied by the clutch pushrod adjuster and check that at least ¼ inch of the bolt has engaged with the internal thread. Turn the nut in a clockwise direction until the diaphragm spring is flat and free to rotate. This will enable the retaining circlip to be prised from its groove inside the periphery of the

6.5 Remove the inspection cap to check camshaft chain tension

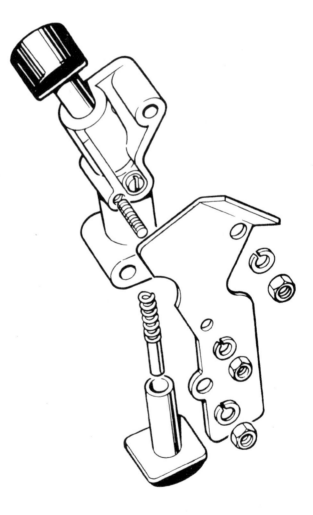

Fig. 9.2 Primary chain hydraulic tensioner

clutch body by a screwdriver blade. Lift the first end clear of the groove, then peel the circlip out of position. The compressor and diaphragm can be lifted away together; there is no necessity to detach the compressor until after the clutch is reassembled. NEVER peel out the circlip without first using a compressor to compress the diaphragm.

17 Lift out the clutch plates, using two pieces of stout wire with their ends bent at a right angle to form a hook. There is a total of eight, four plain and four friction. The clutch inner drum can now be released by unscrewing the centre nut . To prevent the clutch from turning, select top gear and apply the rear brake by pressing on the operating arm. The centre nut has a right hand thread and should be removed, after bending back the tab washer, complete with the spring washer beneath it. Pull out the clutch pushrod. The clutch inner drum cannot be withdrawn.

18 The triplex primary chain is of the endless type and has no split link joint. In consequence it is necessary to remove the clutch outer drum with its integral sprocket and the engine sprocket in unison. The engine sprocket is a keyed taper fit on the end of the crankshaft and it is essential to use a sprocket puller to achieve its release. Norton Villiers service tool 060941 is specified for this purpose; the sprocket is tapped to accept the extractor bolts. If the service tool is not available, a two or three legged sprocket puller can be used with equal effect. Lift away both sprockets together with the triplex chain. Take special care of the collar and spacers fitted over the gearbox mainshaft, behind the clutch, since they determine the accurate alignment of the two sprockets. Place them in a safe place until reassembly commences.   There are spacers on the crankshaft, behind the engine sprocket location, and a Woodruff key.

19 Remove' the clutch adjustment circlip from the gearbox mainshaft. Remove the inner chaincase retaining nut from the central boss. The inner chaincase cover is now free to be pulled off its studs, complete with the electric starter motor.

5 Dismantling the engine - removing the crankcase assembly from the frame

1   Unscrew the tachometer cable from the union joint in front of the right-hand cylinder barrel. At the same time it is convenient to drain the oil from the tank by removing the drain plug, or in the case of the early models, by taking out the larger oil filter union which secures the main oil feed pipe from the oil tank. It will be necessary to detach the right hand side cover in order to gain access to the oil tank. This task is best accomplished whilst the oil is warm, so that it will flow more freely.

2   Remove the rocker feed pipe which is joined to the timing cover by a banjo union. Take care not to lose the copper sealing washers.

3   Remove the neutral light lead at its snap connector and disconnect the earth lead, located at the rear of the crankcase (if it has not previously been unbolted).

4   Slacken the Jubilee clip and remove the timing chest breather pipe.

5   Place a large capacity tray under the crankcase assembly and remove the oil pipe junction block from the right hand crankcase, assuming the oil tank has drained completely. Slacken and remove the hexagon headed drain plug in the bottom of the left-hand crankcase (7/8 inch Whitworth). Allow all excess oil to drain off.

6   It is necessary to remove the crankcase assembly by detaching the front engine mounting. Remove the large diameter bolt which passes through the centre of the mounting, taking care to align the flats on the head so that the bolt will clear the timing case during removal. The Mark 3 Isolastic mountings are adjustable and the new design allows them to be removed in one piece, with the end caps in position.

7   Remove the two nuts from the right hand side (timing side) of the engine mounting studs and pull the stud out, complete with the remaining nuts, from the left-hand side. The front mounting is now free to be removed from the crankcase and frame.

8   Remove the bottom rear crankcase/engine plate bolt and the bottom centre stud. Remove the stud by lifting the front of the crankcase to obtain the necessary clearance from the lower frame tubes. It is advisable to have the drain tray in position, since as the drain plug is removed some additional oil will escape from the crankcase as its position is altered.

9   Support the crankcase by placing a metal bar between the sump plug and lower front crankcase bolt and remove the upper rear nut. Pull out the bolt. The crankcase assembly is now free to be removed from the right-hand side of the machine. Before removing , check that all the necessary components eg., oil pipes, electric cables etc., have been disconnected.

6 Timing cover anti-drain valve and priming the oil pump

1   Mark 3 models are fitted with an anti-drain valve to prevent oil seepage past the pump when the engine is left standing. The plunger and spring are located in the timing cover where the conical rubber oil seal of the oil pump makes contact.

2   Check that the plunger moves freely in the timing chest, before reassembly.

3   Due to the presence of the anti-drain valve, the oil pump now requires priming (if it has been removed) before the engine is started. Prime the pump by turning the pump drive shaft by hand, whilst pumping oil into the gears from an oil can. Also, after replacing the crankcase in the frame and connecting up the oil pipes, fill the oil tank and slacken the oil crankcase junction block screw. When oil issues from the gap, torque tighten the screw to 8lbs/ft (1.10kg/m). The oil pump and feed pipes are now primed. **IT IS IMPERATIVE TO PRIME THE OIL PUMP AND PIPES BEFORE STARTING THE ENGINE** after a rebuild, otherwise there is every possibility of a seizure.

4   On Mark 3 models all the timing cover retaining screws are now the same length.

5   An inspection plug is fitted to the timing chest to permit visual inspection of the camshaft chain tension. The timing cover has, however, to be removed in order to adjust the chain tension. Removal of the cover is always preferable when checking the tension since the valve spring pressure must be relieved from the crankshaft by holding the camshaft sprocket nut in the appropriate position ie; no tension in the chain.

7 Crankcase and main bearings

1   On Mark 3 models the drive side oil seal is now retained by a circlip.

2   Both main bearings are of the roller pattern (cf. a ball journal bearing on the right-hand side on earlier models). The timing side crankcase half can now be lifted away from the crankshaft assembly without difficulty.

8 Crankshaft, big end and engine bearings - examination and renovation

1   Check the big end bearings for wear by pulling and pushing on each connecting rod in turn, whilst holding the rod under test in the vertical plane. Although a small part of side play is permissible, there should be no play whatsoever in the vertical direction if the bearing concerned is fit for further service.

2   The big end bearings take the form of shells. To gain access to the shells, remove the connecting rods by unscrewing the two self-locking nuts at each end cap. When the nuts have been withdrawn completely, the connecting rod and end cap can be pulled off the crankshaft, with the bearing shells still attached. Mark both the connecting rods and their end caps clearly so that there is no possibility of them being interchanged. Note that the locating tabs of the bearing shells fit to the same side of each connecting rod.

3   If the crankshaft assembly is to be separated, it is advisable to

ALL NUTS NOW $\frac{3}{8}$ = 14 OFF

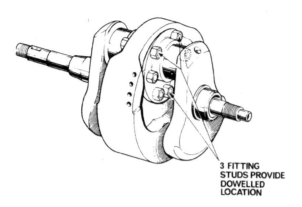

3 FITTING
STUDS PROVIDE
DOWELLED
LOCATION

Fig. 9.3 Position of crankshaft locating studs

continue the dismantling in a clean metal tray. The assembly holds approximately one teacup of oil which will be released when the cranks are separated from the centre flywheel. Mark both the crankcheeks and centre flywheels so that they can be aligned correctly on reassembly. Also mark the flywheel, to prevent it from being reversed on reassembly.

4   Remove the seven nuts on the right hand (timing) side of the crankshaft assembly. All will be tight.

5   The Mark 3 crankshaft is of a stiffer construction than previous assemblies and the drive side shaft has been extended to accommodate the electric starter mechanism. The diameter of the crankshaft studs has been increased to 3/8 in. On early Mark 3 models, crankshaft alignment was obtained by one 7/16 in./3/8 in. shouldered stud on the crank centre line and two 3/8 in. diameter studs in the 11 and 1 o'clock positions. The shouldered stud has however been replaced by a parallel 3/8 in. stud on the latest crankshafts.

6   Jar the crank flanges from the flywheel with a hammer and a soft metal drift. This will release the left hand (drive side) crankshaft. If the machine has covered a considerable mileage, it will probably be found that there is a build up of sludge in both crank flanges and in the flywheel recess as a result of the centrifugal action of the rotating assembly. This must be cleaned out thoroughly prior to reassembly.

7   Wash each crank in turn with clean petrol and use compressed air to dry off. Light score marks on the big end journals can be removed by the use of fine emery cloth but if the scoring is excessive or deep, or if measurements show ovality of more than 0.0015 in. (0.0381 mm) the journals must be reground. The accompanying illustration shows the regrind sizes permissible; shell bearings are available in undersizes to match, from minus 0.010 in. to 0.040 in. in 0.010 in. stages. The big end shells are finished to give the required diametrical clearance and must not be scraped to improve the fit.

8   Main bearing roller races should be withdrawn from the crankshaft by using Norton Villiers service tool 063970. There is insufficient clearance for the legs of a sprocket puller or other substitute, if the service tool is not available. Main bearing failure is characterised by a rumbling noise from the engine and some vibration which may not be damped out by the Isolastic engine mountings. Bearing of the ball or roller type should be renewed if any play is evident, if the tracks are worn or pitted, or if any roughness is felt when they are rotated by hand

9   Engine reassembly - rebuilding the crankshaft assembly

1   Before the crankshaft is reassembled, check that all parts are clean and that the oilways are free, preferably by blowing through with compressed air. Arrange the parts in the correct order for reassembly, taking note of the alignment marks made when the crankshaft was dismantled.

2   Fit the left-hand (drive side) crank cheek to the flywheel aligning the marks made when dismantling took place. Fit the central dowel stud through from the left-hand to the right-hand side followed by the remaining two dowel studs that are in the 11 and 1 o'clock positions.

3   Replace the right-hand (timing side) crank cheek over the three studs. Tap it home with a soft face hammer. Replace the four remaining studs.

4   Apply Locktite LT241 or a similar thread locking compound to the stud threads. Replace and tighten the nuts in a diagonal sequence to a torque wrench setting of 30lbs/ft (4.15kg/m)

5   Check to ensure the oilways blanking plug is fitted in the right hand (timing side) crank cheek. This is important if a new right hand (timing side) crank cheek has been fitted. Centrepunch the nuts as a safeguard against slackening.

6   Pump oil through the crankshaft assembly with a pressure oil can to ensure all the oilways are clear and that the oil flow is not impeded in any way. A considerable amount of oil will be needed before it excludes from the oilways, due to the need to fill the area within the flywheel centre and both crank flanges.

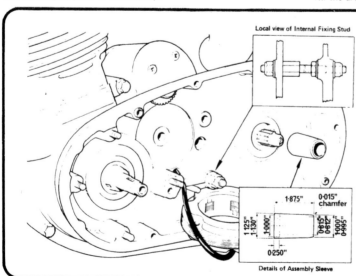

Local view of Internal Fixing Stud

Fig. 9.4 Replacing the primary chaincase inner cover

Details of Assembly Sleeve

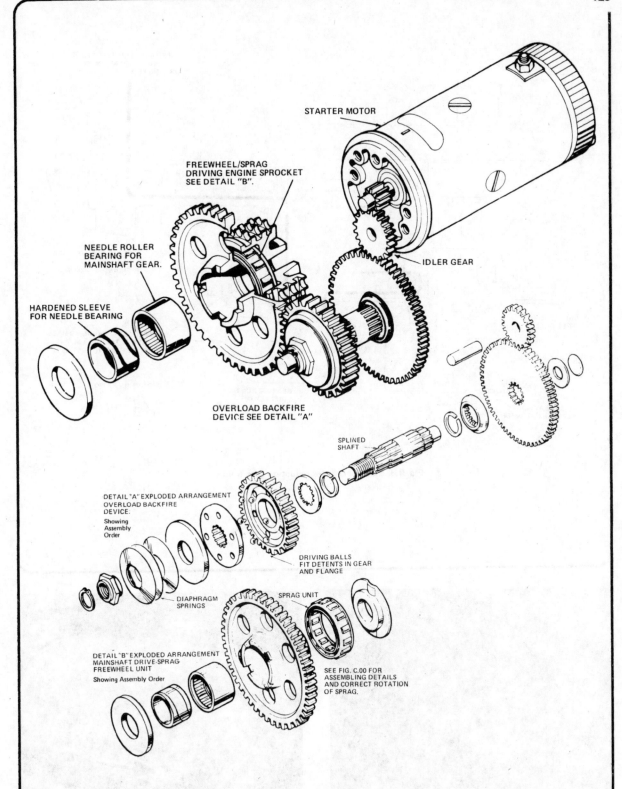

STARTER MOTOR

FREEWHEEL/SPRAG
DRIVING ENGINE SPROCKET
SEE DETAIL "B".

NEEDLE ROLLER
BEARING FOR
MAINSHAFT GEAR.

IDLER GEAR

HARDENED SLEEVE
FOR NEEDLE BEARING

OVERLOAD BACKFIRE
DEVICE SEE DETAIL "A"

SPLINED
SHAFT

DETAIL "A" EXPLODED ARRANGEMENT
OVERLOAD BACKFIRE
DEVICE.
Showing
Assembly
Order

DRIVING BALLS
FIT DETENTS IN GEAR
AND FLANGE

DIAPHRAGM
SPRINGS

SPRAG UNIT

DETAIL "B" EXPLODED ARRANGEMENT
MAINSHAFT DRIVE-SPRAG
FREEWHEEL UNIT

Showing Assembly Order

SEE FIG. C.00 FOR
ASSEMBLING DETAILS
AND CORRECT ROTATION
OF SPRAG.

Fig. 9.5 Electrical starter - general arrangement

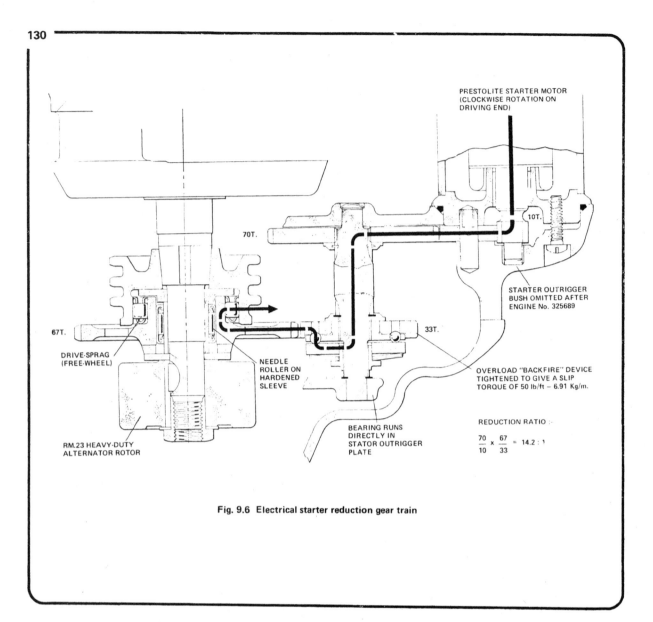

PRESTOLITE STARTER MOTOR
(CLOCKWISE ROTATION ON
DRIVING END)

70T.

10T.

67T.

DRIVE-SPRAG
(FREE-WHEEL)

NEEDLE
ROLLER ON
HARDENED
SLEEVE

33T.

STARTER OUTRIGGER
BUSH OMITTED AFTER
ENGINE No. 325689

OVERLOAD "BACKFIRE" DEVICE
TIGHTENED TO GIVE A SLIP
TORQUE OF 50 lb/ft – 6.91 Kg/m.

BEARING RUNS
DIRECTLY IN
STATOR OUTRIGGER
PLATE

RM.23 HEAVY-DUTY
ALTERNATOR ROTOR

REDUCTION RATIO :-

$$\frac{70}{10} \times \frac{67}{33} = 14.2 : 1$$

Fig. 9.6 Electrical starter reduction gear train

11.13 Tighten the chain tensioner nuts to the correct torque
setting

11.23 Fill the chaincase up to the oil level screw

10 Backfire overload device

1 The backfire overload device should not be dismantled since it has been preset by the factory

2 If the device has been dismantled, it must be preloaded on assembly to slip at a torque wrench setting of 50lbs/ft ± 2.0 lbs/ft (6.92kg/m ± 0.28 kg/m). The outer edge of the adjuster nut must be peened over to prevent movement.

11 Engine reassembly - reassembling the primary transmission and electric starter mechanism

1 Replace the starter motor and small intermediate gear in the inner primary chaincase. Do not forget to fit the large rubber O-ring. Tighten the two screws evenly.

2 Fully screw the backing nut onto the central supporting nut, then replace the washer. Replace the Woodruff key used to locate the engine sprocket on the left-hand end of the crankshaft. Lightly smear the smooth circular face of the left hand crankcase with gasket cement and the matching face of the back of the inner chaincase casting. Fit a new gasket to the crankcase face and fit the inner chaincase, taking care that the gearbox mainshaft does not damage the oil seal within the centre of the rear of the casting. It is advisable to grease both the seal and the mainshaft, or cover the splines with plastic tape, to obviate risk of damage. To prevent damage to the oil seal, as assembly sleeve is available (part No. ST4928D) which can also be turned up on a lathe.

3 Check that the chaincase is aligned correctly. Tighten the backing nut on the central supporting stud until it touches the chaincase. Replace and tighten the retaining nut and washer (see Fig. 9.4).

4 Replace the clutch locating circlip and slide the clutch location spacer over the gearbox mainshaft, recessed portion inwards. Add the spacing washers, necessary to ensure correct alignment of the sprockets and chain.

5 The triplex chain has no spring link and it is necessary to fit the engine sprocket, chain and clutch as a unit. Before assembling and fitting these components, check the centre bearing of the clutch. This is a ball journal bearing, retained by a circlip. If any play is evident or if the bearing runs roughly as the clutch body is rotated, it should be replaced. It is a drive fit in the clutch body.

6 With the large starter gear located in the machined recess behind the top run of the primary chain, fit the clutch, engine sprocket and chain over the respective shafts. The clutch has a splined centre and the engine sprocket has a keyway which engages with the key already inserted in the crankshaft. If necessary, use a tubular drift to ensure both sprockets are correctly located to the full depth of engagement. Fit the clutch centre securing nut and tab washer. Lock the engine by either of the methods described in this Chapter; replace and torque tighten the clutch centre nut to 70 lbs/ft (9.678 kg/m). Bend over the lockwasher onto two flats of the nut.

7 Reassemble the clutch plates. The plain steel plate with the two small dowels on the inner face must be fitted first; the dowels engage with matching holes in the main body of the clutch. Then fit the friction and steel plates in alternate order, ending with the extra thick steel plate which has a serrated outer rim. Later models may have some variation of the clutch make-up including the use of sintered bronze friction plates in place of the earlier friction material used and a specially-hardened clutch centre. The method of assembly is, however, broadly identical.

8 The clutch diaphragm should be reassembled with the compressor tool and tensioned so that it is completely flat. Push the diaphragm, complete with compressor tool, as far into the clutch as possible and enter one end of the retaining circlip into the groove within the clutch body. Feed the remainder of the circlip into the groove and check that it has located correctly

BEFORE RELEASING THE COMPRESSOR. This precaution cannot be overstressed. If an attempt is made to fit the diaphragm without the correct type of compressor, or if the retaining circlip is not located positively, THERE IS RISK OF PERSONAL INJURY if the compressed diaphragm works free.

9 Replace the clutch pushrod within the hollow mainshaft, after coating it with grease. Insert it through the centre of the clutch, and replace the pushrod adjuster screw and locknut. Adjust the pushrod by slackening off the handlebar adjuster completely and screwing in the adjuster until the clutch commences to lift. Slacken back the adjuster one complete turn and lock it in this position with the locknut. It may be necessary to detach the inspection cover from the gearbox during this operation because if the clutch operating arm within the gearbox outer shell has dropped out of location, clutch action will be rendered inoperative. It can be raised back into position if the clutch adjuster is temporarily slackened off.

10 Replace the thrust collar in the sprocket, with the small diameter end facing outwards. Slide on the hardened sleeve, using a tubular drift if it is a tight fit.

11 Replace the sprag clutch in the sprocket. It is imperative that it is fitted the correct way round ie; sprags leaning to the left when viewed at the top of the sprocket (see Fig. 9.7).

12 Replace the crankshaft starter gear complete with needle roller bearing, over the hardened sleeve. Revolve the gear anticlockwise to check that the sprags are driving the correct way to start the engine.

13 Assemble the chain tensioner, making sure the plungers are in their correct positions. Replace the tensioner over the studs. Apply a thread locking compound to the studs and torque tighten the 5/16 in. nuts to 12 lbs/ft (1.66 kg/m) and the ¼ in. nut to 4-5 lbs/ft (0.553 - 0.691 kg/m). These torque figures are critical since the plungers are likely to jam if the nuts are over-tightened.

14 Prime the chain tensioner by squirting a little oil into the tensioner housing.

15 Replace the backfire device on its spline. Fit the outrigger plate over the four studs whilst locating the backfire device's shaft into the bearing. Replace the lockwashers and nuts. Bend the tabs on the lockwashers backwards over the outrigger plate. Torque tighten the nuts to 15 lbs/ft (2.07 kg/m). Knock up the other tabs of the lockwashers onto the flats of the nuts.

16 Replace the large thrust plate on the mainshaft and fit the Woodruff key. Clean the rotor of any metal particles that may have been attracted to it. Replace the rotor with its name and timing marks facing outwards. Replace the shaped washer and torque tighten the rotor nut to 70 lbs/ft (9.68 kg/m).

17 Replace the stator on the three studs, with the leads facing outwards and positioned at 5 o'clock. Pass the leads through the rubber grommet in the inner chaincase. Reconnect the leads at their snap connectors.

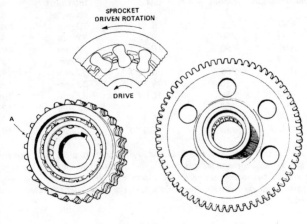

Fig. 9.7 Assembly position of the sprags

18 Replace the plain washers and stator retaining nuts. Tighten the nuts to a torque setting of 15 lbs/ft (2.07 kg/m). There must be a minimum air gap between the stator coil assembly and the rotor of from 0.008 - 0.010 inch and a check should be made with a feeler gauge. If the gap is reduced at any point, misalignment of the stator mounting studs should be suspected and corrected.

19 Pass the gearbox crossover shaft, complete with pinion, through the oil seal in the inner chaincase. Locate the splined end with the coupler, which should be fitted to the gearbox.

20 Fit a new chaincase gasket, holding it in position with a little gasket cement. Offer up the outer chaincase cover complete with gear lever and pinion. Place the gear lever in the required position, engage the two pinions and fit the chaincase. Replace the eleven screws; the long screws securing the electric starter. Tighten the screws evenly and in a diagonal sequence.

21 Refit the left-hand footrest with the red earth wire terminal from the zener diode attached to the rear bolt.

22 Replace the starter motor cable, and the earth wire at the rear of the crankcase.

23 Refill the primary chaincase with oil until it just begins to run out of the oil level screw (approx. 300 cc). Note: The primary chaincase oil level is critical. If there is insufficient lubricant it will not be thrown off the clutch sprocket into the chain tensioner and thus the tensioner will become inoperative.

12 Engine reassembly - head steady and suspensory spring device

1 When tightening the head steady make sure that the machine is standing on its wheels and not on the centre stand. This procedure ensures that the rubber bushes are not preloaded.

2 When replacing the suspensory spring device, check the spring setting load (see Fig. 9.8) and adjust as necessary by turning the self locking nut in the appropriate direction to obtain the correct figures. The adjuster nut can then be adjusted by 6 flats at a time between test runs until optimum smoothness is obtained.

13 Increasing engine performance - cylinder head identification and compatability

1 Cylinder heads are identified by a part number and code stamped on the right exhaust rocker box face. Below are the relevant details.

14 Gearbox - dismantling and reassembly

The left-hand gearchange mechanism has not altered the basic dismantling and reassembly procedures. The relevant gearbox design changes are detailed in the following Section and should be noted before undertaking any gearbox maintenance.

15 Gearbox - general

1 Although the gearbox remains basically the same as for the previous right-hand gearchange models, there are some minor changes which are covered in the points below.

2 A neutral light switch has been fitted, the adjustment of which is covered in Section 16 of this Chapter.

3 To obtain clearance for the left-hand gearchange crossover shaft, the shifter quadrant is now retained by a circlip instead of a bolt and washer.

4 A small brass breather has been fitted to the top of the gearbox; this supersedes the breather hole that was provided in the inspection cover.

5 The knuckle pin roller in the shifter quadrant has been replaced by a spherical trunnion, which means that it can be

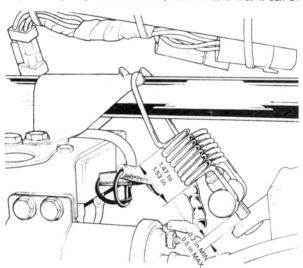

Fig. 9.8 Suspensory spring device; fitting dimensions

Identification	Model	C.R.	Inlet port diameter	Remarks
RH1	750	9 : 1	30 mm	Standard up to 1972
RH2	750	10.25	32 mm	AMA Racer
RH3	750	10 : 1	32 mm	1972 Combat
RH4	850	8.5 : 1	32 mm	1973 850 Model
RH5	750	8.9 : 1	32 mm	1973 low compression ratio
RH6 *	750	9.3 : 1	32 mm	1973 (preferred stock item)
RH7	750	10 : 1	32 mm	1973 Short stroke (Standard)
RH8	750	–	32 mm	1973 Short stroke (Race kit) NS
RH9	850	10.5 : 1	32 mm	1974 high performance
RH10	850	8.5 : 1	30 mm	1974 850 model

* Adjust C.R. using cylinder base and head gaskets (see NVT Service Release N3/21 and N3/23).
Tapered carburettor adaptors and parallel carburettor adaptors are available. See relevant parts lists.
It should be noted that early 750 model cylinder heads are now supplied fitted with 850 model (cast iron) valve guides. These heads continue to be identified under the original R.H. identification, but additionally are stamped with the suffix (S) after the number RH6S).

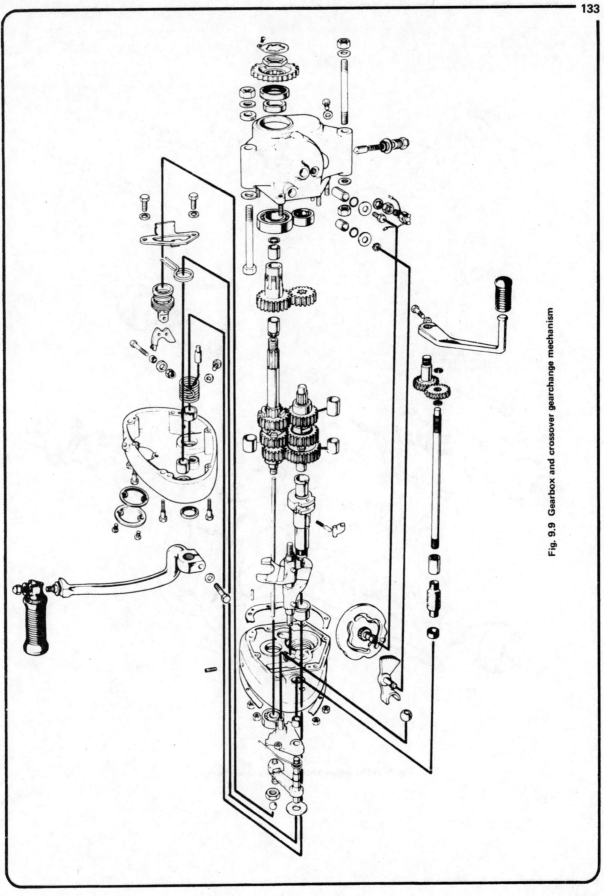

Fig. 9.9 Gearbox and crossover gearchange mechanism

134

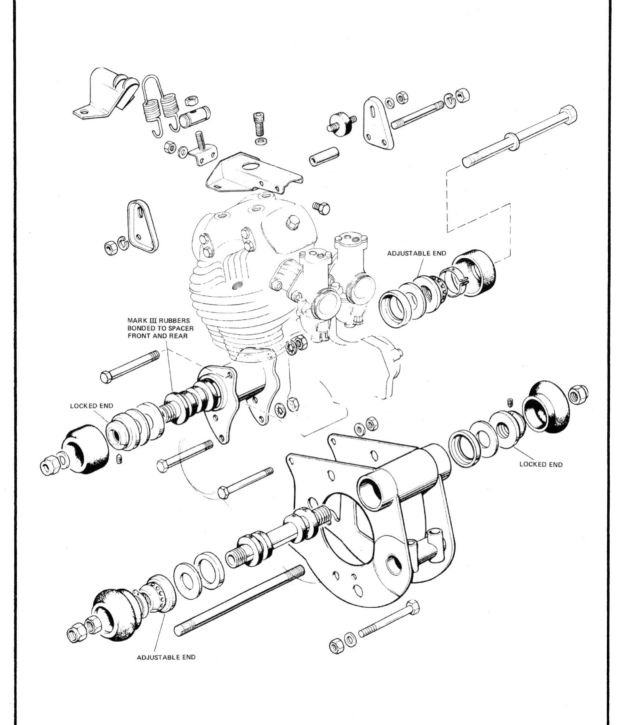

ADJUSTABLE END

MARK III RUBBERS
BONDED TO SPACER
FRONT AND REAR

LOCKED END

LOCKED END

ADJUSTABLE END

Fig. 9.10 The adjustable Isolastic mountings

removed and replaced without having to detach the gearbox inner cover.

6   The indexing plunger domed nut is now fitted with a rubber O-ring and washer.

7   On the latest Mark 3 models the second gear ratio has been changed. Both the mainshaft and layshaft second gears have been modified, the new teeth numbers are 18 and 23 respectively (instead of 18 and 24). They are, therefore, only replaceable as a pair since the pitch is different. The new gears are identified by a groove cut in the ends of the dogs.

8   The gearbox outer cover no longer has a hole for the gearchange lever shaft. The shaft now extends from the forward left-hand side of the inner cover, via an oil seal. A splined coupler in a rubber sleeve is used to connect it to the crossover shaft. Both shafts are a sliding fit in the coupler. Excessive entry of either shaft is prevented by a circlip fitted in the middle of the coupler.

16 Neutral light switch - adjustment

1   The neutral light switch is located in the lower, forward-facing part of the gearbox shell. It is actuated by a domed blip on the camplate which presses in the operating ball of the switch and thereby closes the switch contacts. If the switch is faulty it will have to be renewed since it is a sealed unit.

2   To adjust the switch (or if the switch has been removed completely) first select neutral. Slacken the locknut and screw in the switch body until the neutral indicator light comes on. Tighten the locknut.

3   Change gear a few times, then re select neutral to make sure that the light still comes on. If not, reset by screwing the switch body further into the gearbox case. Do not screw the switch further in than necessary, otherwise it will be damaged.

17 Ignition timing - setting

1   To facilitate ignition timing a slot has been machined in the crankshaft cheek. This slot should align with the centre of the aperture left when the inspection plug has been removed. The plug is located in the front underside of the timing chest. This positioning gives the initial ignition timing setting of approximately 28º BTDC. Otherwise, proceed as for earlier models.

18 Front forks and brake

1   On Mark 3 models the disc brake assembly has been relocated on the left-hand side. The fork legs have been transposed, therefore all previous procedures for dismantling, examination, renovation and reassembly will apply.

19 Isolastic engine mountings - adjustment

1   The Isolastic engine mountings have been modified so that they can be adjusted without need for re-shimming. The rubbers are no longer individually replaceable since they are moulded onto the sleeve. To adjust, follow the procedure given below.

2   Slacken the through engine bolt nut by at least two turns.

3   Slide off the protective clip on the adjuster ring, and tighten the adjuster ring to take up all the clearance by turning in a clockwise direction. Use either the special tool provided or a small steel bar.

4   Back the adjuster ring off 1/5 turn (1½ holes) and torque tighten the through bolt nut to 30 lb/ft (4.15 kg/m). Do not forget to replace the protective clip. By backing off the adjuster ring a clearance of 0.006 in. (0.152 mm) should be obtained. **Note:** Mark 3 models are fitted with bronze loaded PTFE

17.1 Remove the plug to set the ignition timing

18.1 The front brake has been transposed to the left-hand side

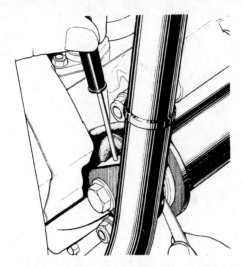

Fig. 9.11 Checking the Isolastic engine mountings

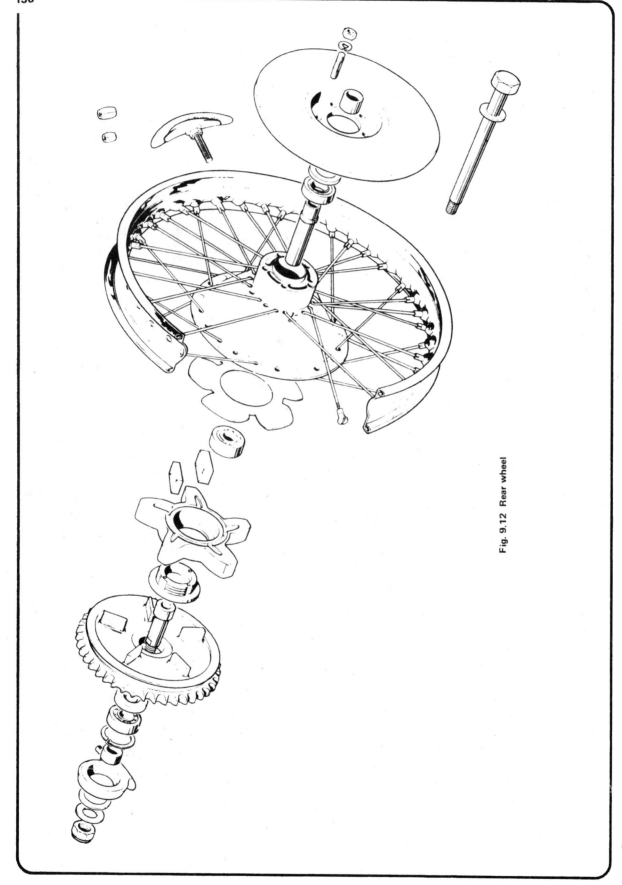

Fig. 9.12 Rear wheel

washers in place of the original cream coloured polyurethane washers. These new washers are more wear resistant and should be fitted on earlier models when replacements are required.

20 Front wheel bearings - removal and replacement

1 The threaded locking ring has been superseded by a circlip; otherwise the procedure remains the same.

21 Rear wheel - examination, removal and renovation

1 The rear wheel on the Mark 3 models is completely new and is fitted with a disc brake. The speedometer drive gearbox is now located on the left-hand side.
2 Before removing the rear wheel, check for rim alignment, damage to the rim and loose or broken spokes by following the procedure described for the rear wheel, see Chapter 7 Section 2.
3 To remove the rear wheel without disturbing the final drive chain, place the machine on its centre stand so that the rear wheel is clear of the ground. Engage either first or second gear to simplify refitting.
4 Slacken the lower right-hand suspension unit nut and withdraw as far as the locating circlip.
5 Unscrew and remove the wheel spindle from the right-hand side. Tilt the wheel and remove the chain tensioner.
6 Carefully lift up the caliper complete with its mounting plate but do not remove the caliper completely from the disc. Squeeze together the chain tensioner arms and insert it between the friction pads to prevent them from falling out. The caliper can now be lifted clear of the disc and hung from the hook provided on the frame top loop (early models are not fitted with a hook, therefore use a piece of wire to suspend the caliper).
7 To provide the necessary clearance to remove the rear wheel lean the machine to the left.
8 To replace the rear wheel, reverse the above procedure. Before using the machine, pump the rear brake pedal hard to realign the pads with the disc.

22 Rear wheel - dismantling, examining and reassembling the hub

1 Remove the rear wheel as described in the preceeding Section.
2 Remove the disc by undoing the 5 nuts.
3 If possible pre heat the hub to a maximum of 100°C (pour

boiling water over it) to facilitate the removal of the bearings.
4 Unscrew the lock ring on the cush drive side of the hub. **Note**; dependent on the year of manufacture the ring may have either a right-hand or left-hand thread. Experiment carefully to determine the thread type and then remove accordingly. Use a peg spanner (service tool 063965) or alternatively a punch and hammer
5 Insert the wheel spindle through the hub from the disc side. Using a soft faced hammer, drift in the disc side bearing which, in turn, will push out the left-hand bearing.
6 To remove the larger right-hand bearing again, use the wheel spindle and a suitable spacer. If the bearing spacer tube is used, be careful not to burr over the ends. The oil seal will be drifted out, along with the bearing.
7 Remove the old grease from the hub and bearings. Examine the bearings for wear eg; roughness when they are rotated, or play. Renew if in any doubt.
8 Pre pack the bearings and hub with a high melting point grease. Allow plenty of space in the hub to accommodate the expansion of the grease when it becomes hot.
9 Reassembly is basically a reversal of the dismantling procedure. Replace the cush drive bearing, oilseal side outwards. Tighten the locking ring. When replacing the bearings use a tubular drift (a socket can often be used) bearing only on the outer race to prevent damage.
10 Replace the bearing spacer tube until it "sounds" fully home. Renew and replace the oil seal.
11 Refit the disc and nuts, do not forget the spring washers. Torque tighten the nuts evenly and diagonally to 20 lbs/ft (2.76 kg/m).

23 Rear sprocket, speedometer drive gearbox and cush drive - removal, examination and replacement

1 Remove the rear wheel as described in Section 21 of this Chapter.
2 Disconnect the chain at its spring link.
3 Disconnect the speedometer cable from the speedometer drive gearbox now on the left-hand side of the machine.
4 Undo the rear sprocket spindle nut and lift the sprocket out of the swinging arm. Lift off the speedometer drive gearbox, note the spacing collar.
5 Remove the sprocket spindle and extract the circlip. Drive out the bearing from the seal side and then drift out the seal. To ease removal pre heat the carrier to a maximum temperature of 100°C. Be careful not to damage the speedometer drive slots when using the drift.

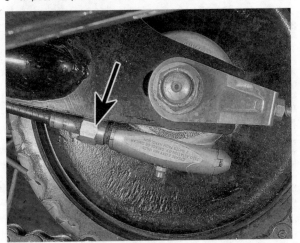

23.3 Disconnect the speedometer cable

24.1 The rear brake assembly

6   Remove all the old grease and examine the bearing for wear eg; roughness when rotated or play. Renew, if in any doubt. Also check that the sprocket teeth are not worn or hooked. If a replacement is required, it is good policy to also renew the chain and gearbox sprockets, since they are also likely to be worn.

7   To replace, pack the new bearing with high melting point grease and drift it into position in the sprocket. Use a tubular drift, bearing on the outer race only to prevent damage (a socket can often be used).

8   Replace the circlip, sharp edge outwards. Drift in a new oil seal with the metal flange away from the bearing.

9   Slide in the sprocket spindle. Replace the spacing collar and the speedometer drive gearbox, complete with its spacing washer. Pre lubricate the gearbox with a little high melting point grease.

10 Replace the sprocket unit in the swinging arm; do not forget to fit the chain adjuster.

11 Before replacing the rear wheel, check the condition of the cush drive rubber in the rear wheel hub. Renew if there is any signs of deterioration.

12 Replace the rear wheel by reversing the removal procedure.

24 Rear brake - caliper and disc

1   The caliper and disc are of identical design to that fitted to the front wheel, thus the information given for the front is also applicable to the rear.

25 Rear brake pedal - adjustment

1   The rear brake pedal can be adjusted for height by altering the effective length of the pushrod.

2   Slacken the locknut that abuts the clevis pin (nut A. see Fig. 9.13). Turn the pushrod by nut B in a clockwise direction (when viewed from the rear of the master cylinder), to lower the pedal or nut C anti-clockwise to raise the pedal. Do not forget to re-tighten the locknut A. **Note:** On no account try to separate nuts B and C from each other. They have been locked together during manufacture and disturbing them will alter the piston stroke which could result in the rear brake becoming inoperative.

26 Brake light switch - replacement

1   Identical switches are used for the front and rear brake light. If the brake light fails, first check the wiring and bulb before renewing the switch. The switch is a sealed unit and cannot be repaired if it is found to be faulty.

2   To remove the switch, pull back the plastic cover and disconnect the spade terminal connectors. The switch screws into the master cylinder. Have the replacement switch handy to prevent excessive loss of brake fluid. Be extremely careful not to allow any brake fluid to contact the paintwork. If by accident it does, flush off with plenty of water to prevent the paintwork blistering.

27 Electrical system - specification where different from previous models

Battery	13 or 14 amp/hr @ 10 hour rate	
Alternator	Manufacturer	Lucas
	Type	RM23
	Rate	180 watts
Rectifier	Manufacturer	Lucas
	Type	2DV.406
Warning light assimilator	Manufacturer	Lucas
	Type	066393 Standard 066392 Canada
Starter motor	Manufacturer	Prestolite
	Type	MGD 4111
Solenoid	Manufacturer	Prestolite
	Type	SAZ 4201N
Brake light switch	Manufacturer	
	Type	2SH (2 off)

28 Electrical system - general

1   The electrical system has been considerably altered with new switch gear, electric starter and a higher output alternator. The alternator now feeds a halfwave rectifying system with twin zener diode regulation. Also new is a solid state encapsulated warning light assimilator.

2   To meet Canadian legislation for headlight illumination when the engine is running, a special master switch and encapsulated device are fitted.

3   The assimilator box is located in the battery box immediately in front of the battery. The new zener diode is fitted to the left-hand alloy footrest support plate; the other diode is in a similar position on the right-hand side. A power take-off socket has also been provided and is located on the right-hand side of the machine.

29 Starter motor - removal and replacement

1   Pull back the rubber cover and disconnect the heavy cable on top of the starter motor.

2   Remove the three Posidrive screws that retain the motor and withdraw the motor. Note the large rubber O-ring. **Note:** An outrigger bearing previously used is no longer "bushed" since engine No. 325689.

3   Replacement is the reversal of the above procedure.

30 Starter motor - dismantling, examination and reassembly

1   Hold the starter motor in a vice fitted with soft jaws. Using a ring spanner or socket remove the two through bolts.

2   Tap the end of the armature, with a soft faced hammer, to dislodge the blind end cover. Note that both end covers and body are marked to aid correct reassembly. Hold the gear end of the armature and pull the end cover off. Note the order of the spring shim and thrust washer.

3   Again with a soft faced hammer, tap the commutator end of the armature and then pull off the drive end cover.

4   Withdraw the armature. The brushes will spring out of their holders. Note the positioning of their wires so that they can be reassembled correctly.

5   Inspect the condition of the commutator; if it is badly scored it will have to be skimmed by an electrical specialist.

6   Check the condition of the bushes and renew if necessary. Preheat the bush housings to facilitate removal and fitting. The blind bush will have to be extracted by first tapping a suitable thread in it and then withdrawing it, using a bolt and spacer.

7   It is advisable to renew the oil seal, particularly if there is any signs of oil in the motor.

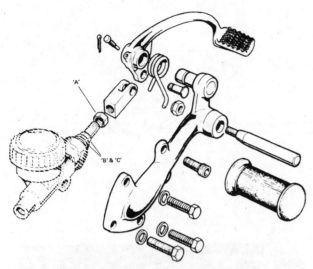

Fig. 9.13 Rear brake pedal and height adjuster

26.2 Pull back the protective cover and disconnect the wires

28.3a The new left-hand zener diode and ....

28.3b The existing right-hand zener diode

28.3c The power take off socket

29.1 Disconnect the power cable to the starter motor

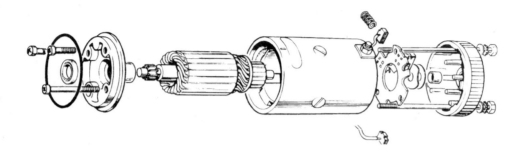

Fig. 9.14 Starter motor

8 Check the brushes for wear (compare with the new part) and
renew as necessary. To renew, cut off the brush that is crimped
to the field winding and unsolder the other one. The new brush
leads will both have to be soldered to their respective positions.
Be careful not to overheat the field insulation. **Note**: Use a high
melting point solder.

9 Reassembly is a reversal of the dismantling procedure. Do not
forget to lightly lubricate the bushes, thrust washer and spring
shim with a little molybdenum disulphide grease. Use a thread
locking compound on the through bolts and tighten them evenly
to 8 lbs/ft (1.10 kg/m).

31 Solenoid - replacement

1 The solenoid is a sealed unit, located underneath the seat, to
the rear of the battery. It can be removed by undoing the two
bolts and disconnecting the electrical leads. If faulty, the unit
cannot be repaired and a replacement will have to be obtained.

31.1 Undo the two bolts to remove the solenoid

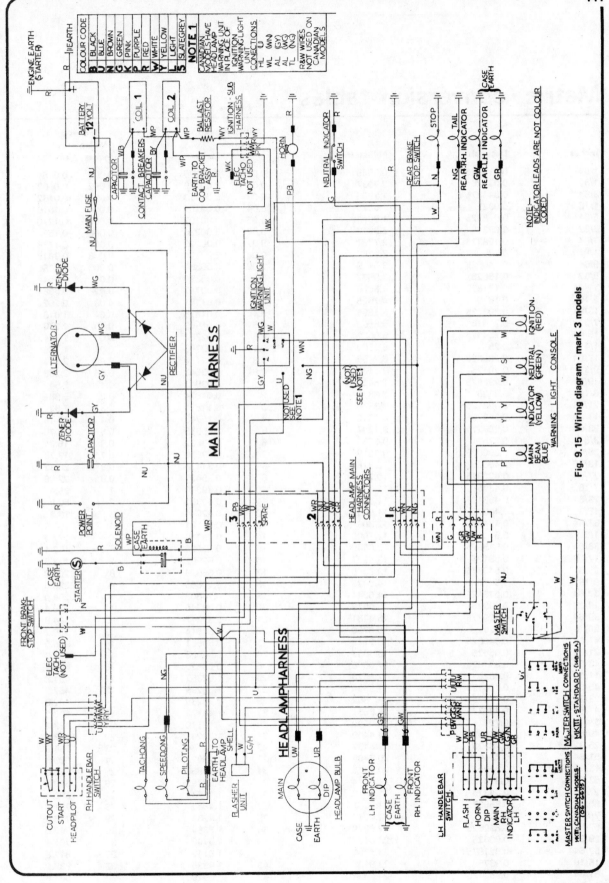

Fig. 9.15 Wiring diagram - mark 3 models

Metric conversion tables

Inches	Decimals	Millimetres	Millimetres to Inches		Inches to Millimetres	
			mm	Inches	Inches	mm
1/64	0.015625	0.3969	0.01	0.00039	0.001	0.0254
1/32	0.03125	0.7937	0.02	0.00079	0.002	0.0508
3/64	0.046875	1.1906	0.03	0.00118	0.003	0.0762
1/16	0.0625	1.5875	0.04	0.00157	0.004	0.1016
5/64	0.078125	1.9844	0.05	0.00197	0.005	0.1270
3/32	0.09375	2.3812	0.06	0.00236	0.006	0.1524
7/64	0.109375	2.7781	0.07	0.00276	0.007	0.1778
1/8	0.125	3.1750	0.08	0.00315	0.008	0.2032
9/64	0.140625	3.5719	0.09	0.00354	0.009	0.2286
5/32	0.15625	3.9687	0.1	0.00394	0.01	0.254
11/64	0.171875	4.3656	0.2	0.00787	0.02	0.508
3/16	0.1875	4.7625	0.3	0.01181	0.03	0.762
13/64	0.203125	5.1594	0.4	0.01575	0.04	1.016
7/32	0.21875	5.5562	0.5	0.01969	0.05	1.270
15/64	0.234375	5.9531	0.6	0.02362	0.06	1.524
1/4	0.25	6.3500	0.7	0.02756	0.07	1.778
17/64	0.265625	6.7469	0.8	0.03150	0.08	2.032
9/32	0.28125	7.1437	0.9	0.03543	0.09	2.286
19/64	0.296875	7.5406	1	0.03947	0.1	2.54
5/16	0.3125	7.9375	2	0.07874	0.2	5.08
21/64	0.328125	8.3344	3	0.11811	0.3	7.62
11/32	0.34375	8.7312	4	0.15748	0.4	10.16
23/64	0.359375	9.1281	5	0.19685	0.5	12.70
3/8	0.375	9.5250	6	0.23622	0.6	15.24
25/64	0.390625	9.9219	7	0.27559	0.7	17.78
13/32	0.40625	10.3187	8	0.31496	0.8	20.32
27/64	0.421875	10.7156	9	0.35433	0.9	22.86
7/16	0.4375	11.1125	10	0.39370	1	25.4
29/64	0.453125	11.5094	11	0.43307	2	50.8
15/32	0.46875	11.9062	12	0.47244	3	76.2
31/64	0.484375	12.3031	13	0.51181	4	101.6
1/2	0.5	12.7000	14	0.55118	5	127.0
33/64	0.515625	13.0969	15	0.59055	6	152.4
17/32	0.53125	13.4937	16	0.62992	7	177.8
35/64	0.546875	13.8906	17	0.66929	8	203.2
9/16	0.5625	14.2875	18	0.70866	9	228.6
37/64	0.578125	14.6844	19	0.74803	10	254.0
19/32	0.59375	15.0812	20	0.78740	11	279.4
39/64	0.609375	15.4781	21	0.82677	12	304.8
5/8	0.625	15.8750	22	0.86614	13	330.2
41/64	0.640625	16.2719	23	0.90551	14	355.6
21/32	0.65625	16.6687	24	0.94488	15	381.0
43/64	0.671875	17.0656	25	0.98425	16	406.4
11/16	0.6875	17.4625	26	1.02362	17	431.8
45/64	0.703125	17.8594	27	1.06299	18	457.2
23/32	0.71875	18.2562	28	1.10236	19	482.6
47/64	0.734375	18.6531	29	1.14173	20	508.0
3/4	0.75	19.0500	30	1.18110	21	533.4
49/64	0.765625	19.4469	31	1.22047	22	558.8
25/32	0.78125	19.8437	32	1.25984	23	584.2
51/64	0.796875	20.2406	33	1.29921	24	609.6
13/16	0.8125	20.6375	34	1.33858	25	635.0
53/64	0.828125	21.0344	35	1.37795	26	660.4
27/32	0.84375	21.4312	36	1.41732	27	685.8
55/64	0.859375	21.8281	37	1.4567	28	711.2
7/8	0.875	22.2250	38	1.4961	29	736.6
57/64	0.890625	22.6219	39	1.5354	30	762.0
29/32	0.90625	23.0187	40	1.5748	31	787.4
59/64	0.921875	23.4156	41	1.6142	32	812.8
15/16	0.9375	23.8125	42	1.6535	33	838.2
61/64	0.953125	24.2094	43	1.6929	34	863.6
31/32	0.96875	24.6062	44	1.7323	35	889.0
63/64	0.984375	25.0031	45	1.7717	36	914.4

Index